KNOWLEDGE AND VALUE

Tulane Studies in Philosophy

VOLUME XXI

Knowledge and Value

ESSAYS IN HONOR OF HAROLD N. LEE

edited by

ANDREW J. RECK

TULANE UNIVERSITY
NEW ORLEANS
1972

MARTINUS NIJHOFF
THE HAGUE
1972

ISBN 90 247 1321 8

PRINTED IN THE NETHERLANDS

CONTENTS

Preface

Monism, Logos, Fire, and Flux: Shannon Du Bose 9

Kant's Theory of Symbolism: John D. Glenn, Jr. 13

Dualisms in William James's *Principles of Psychology;* Andrew J. Reck 23

Charles Peirce and the Firstness of Process: Sandra B. Rosenthal 39

The Meaning of Sameness or Family Resemblance in the Pragmatic Tradition: David L. Miller 51

Pragmatics and Definite Descriptions: John A. Barker 61

Distinguishing Presupposition in Epistemology: Donald S. Lee 85

Verification in Metaphysics: Peter M. Burkholder 101

Gentzen's Cut Elimination Theorem for Non-Logicians: Larry Miller 115

The Aesthetics of Harold N. Lee: A Reconsideration: Louise Roberts 127

On the Phenomenon of Obligation: Edward G. Ballard 139

Positivistic Paths to Value: Robert C. Whittemore 159

Publications by Harold N. Lee 191

PREFACE

Harold N. Lee retired from Tulane University in June 1970. At first the event was too incredible for us to react. Harold N. Lee is a "character" in the best sense of the term. Surely he would never leave us. He was too much an institution for our institution to proceed without him. But he had attained the mandatory retirement age of seventy, as he himself informed us, and we could not refute the calendar. When at last we came to acknowledge the event, we – his colleagues, professional friends, and former students – realized that we wanted to honor him in a manner more permanent than dinners and parties. So the idea of the present collection of essays dawned.

Harold N. Lee taught philosophy at Tulane University for forty-five years. As professor of philosophy at Newcomb College, the undergraduate women's division of Tulane, and head of the Newcomb philosophy department, he carried a heavy burden of teaching and administration. He introduced many of the courses that are now a basic part of the curriculum, such as the courses in aesthetics and symbolic logic. Shortly after World War II he became chairman of the then newly-formed university department of philosophy and played a major role in the establishment of the doctoral program in philosophy. Throughout the decades he also continued his philosophical researches, and he has published and continues to publish important articles and books at regular intervals. His colleagues throughout the South have long recognized his achievements as a philosopher. He has the unique distinction of being past president of both the Southwestern Philosophical Society and the Southern Society for Philosophy and Psychology.

Harold N. Lee's interests in philosophy are broad, and he

has cultivated them with a remarkable depth of scholarship and critical analysis. The essays offered here reflect the range of his interests – in the history of philosophy, in American philosophy, in epistemology, in metaphysics, in logic, in aesthetics, in ethics, and in value theory. We offer these essays in the hope that they do honor him.

Harold N. Lee was co-founder and long-time co-editor of *Tulane Studies in Philosophy*. It is fitting that the present collection of essays be published as a volume in the series.

ANDREW J. RECK
New Orleans, Louisiana

MONISM, LOGOS, FIRE, AND FLUX

SHANNON DUBOSE
Newcomb College, Tulane University

Descartes recognized that once two things have been distinguished, and their difference firmly established, it is uncommonly difficult to grasp their unity.[1] The difficulty of reassembling divided unities is nowhere greater than in the study of ancient thought. One may grant that the division between mind and matter, or between form and its embodiment, belongs to Anaxagoras or perhaps waits for Plato. But since we are accustomed to some such division, how are we to understand Ionian monism?

Aristotle's general approach to Ionian philosophy has been rather consistently followed, despite attacks on his credentials as an historian of philosophy. Even when scholars take pains to remark that in early thought no distinction occurs between matter on the one hand and either form or mind on the other, they tend to suppose with Aristotle that the Ionians concerned themselves with the material cause (*De Partibus* 640 b 5).[2]

[1] R. Descartes, *Oeuvres*, ed. C. Adam and P. Tannery (Paris: Cerf, 1899), III, pp. 690–695.

[2] J. Burnet, *Early Greek Philosophy*, 4th edition (London: Black, 1930). Burnet treats the evidence of Aristotle with reservations (p. 31), and recognizes the ambiguity of the label "matter" (p. vi); he is never the less persuaded that the Ionians were looking for some sort of Urstoff or primary substance (p. 12n). G. S. Kirk, *Heraclitus: The Cosmic Fragments* (Cambridge: Cambridge University Press, 1962), cited hereafter as '*HCF*', appears to follow Burnet's interpretation. W. K. C. Guthrie does so as well in *A History of Greek Philosophy*, I, (Cambridge: Cambridge University Press, 1962), cited hereafter as '*HGP*', but Guthrie defends Aristotle's merits as an historian (pp. 41 ff). H. Cherniss argues to the contrary in *Aristotle's Criticism of Presocratic Philosophy* (Baltimore: The Johns Hopkins University Press, 1935), (Reprinted New York: Octagon 1964). After exhaustive discussion of Aristotle on his predecessors, Cherniss concludes that the Ionians were interested in the problem of change, and that Heraclitus particularly is a process philosopher (p. 382). I believe that Cherniss' emphasis is correct; but that Burnet rightly observed that the Ionians spoke of something "ageless" and "deathless" (p. 10). Consequently I believe that the subject of Ionian speculation is Being in process.

There is a tendency to treat the Ionians as if they were quite familiar with matter, and attended to the problem of describing it, but had not yet discovered mind or immaterial entities. Indeed, one finds a widespread impulse to take rather too literally the spatial metaphor implicit in "division" and "separation", and a consequent treatment of intellectual discovery as if it were a species of successful geographical exploration. Mind and immaterial entities, however, do not simply occupy a new territory where some intrepid traveller makes an eventual landfall. Whatever definition one makes of the form/matter distinction, the two are not separate from each other as Europe is separate from Australia; either continent might be known without the other, but types of being are mutually dependent. Notions of material and immaterial entities depend upon each other in so far as each is defined by contrast or limitation. One may emphasize, as does Anaxagoras, the contrast between inert and active entities. One may, like Plato, emphasize the difference between a formal principle apprehended intellectually, and a perceptible entity. One may distinguish, with Descartes, between what thinks and what is extended. In all these theories, however, each entity is marked off by the limit of the other. The distinction is a logical one, within a given unity. To suppose otherwise is to suppose that the historical development of thought proceeds like Magellan from island to island and from one continent to another, until at last the whole world is circumnavigated. To follow Aristotle in maintaining that the Ionian monists sought to describe the material cause is to impose upon their views a distinction (however important to later philosophers) which is foreign to them. To abandon this distinction is to recognize that, just as Heraclitus is no idealist,[1] so also is it true – and by the same token – he is no materialist.

Heraclitus is not a materialist in either an Aristotelian or a Cartesian sense of the word 'matter': that is, Heraclitus does not attempt to describe a substratum,[2] nor does he refer to

[1] See G. T. W. Patrick, *The Fragments of the Work of Heraclitus of Ephesus on Nature*, translated from the Greek text of Bywater with introduction and critical notes, 1889. Reprinted Chicago: Argonaut, 1969.

[2] G. S. Kirk points out in *HCF* that the identification of fire as substratum in *De Caelo* 298 b 35 is "sheer Aristotelianism" (p. 17); but Kirk nevertheless goes

an entity characterized by geometrical attributes, moving mechanically. Furthermore, Heraclitus does not, like a Platonist, distinguish between entities which are known by sense, and those which are conceived.

The central considerations for an exposition of Heraclitus' theory are the nature of the Logos, the special status of fire, and the authenticity of the flux doctrine. Abandoning the assumption that reason and nature belong to different categories will permit clarification of the Logos doctrine and the status of fire. This clarification will be further illuminated by Diogenes of Apollonia's defense of monism. This discussion will support the authenticity of the flux doctrine, and vindicate Plato's interpretation of Heraclitus.

Aristotle berated Heraclitus for denying the law of contradiction[1] on the grounds that such denial makes anything true. But Aristotle's criticism comes from a point of view which identifies individuals within species as the real existences (*De Partibus* 644 a 24), and which emphasizes logical analysis and distinction. Heraclitus, by contrast, makes a claim that he "distinguishes each thing according to its nature" (fr. 1),[2] even while he urges that "the Logos is common" (fr. 2). The teaching of the Logos is just that "all things are one" (fr. 50). This statement of the Logos doctrine is not isolated. It is echoed in fr. 10 and again in Heraclitus' doctrine of opposites. Fr. 10 asserts that "things taken together are whole and not whole ... from all things can be made a unity, and from a unity all things." The analytic approach is not supreme; one may also and (if we read this passage in the light of fr. 50) more properly take the synoptic view to recognize the whole as prior. Furthermore, the several assertions of the unity of opposites show, as Kirk (*HCF*, 166) justly observes, that Heraclitus perceived the fundamental continuity of such complementary pairs as hot and cold (fr. 126), youth and age (fr. 88), feast and famine (fr. 67), up and down (fr. 60); and,

on to identify fire as the "originative world mass" (p. 306), which is an analogous interpretation.

[1] *Topics* 159 b 30, *Metaphysics* 1012 a 33; the latter concerns fr. 49a, which Kirk queries (*HCF*, 374).

[2] Fragments are identified here according to the numbering of H. Diels and W. Kranz, *Die Fragmente der Vorsokratiker*, 6th edition (Berlin: Weidman, 1951).

of course, strife and justice (fr. 80).[1] The difference of degree, direction, or aspect which characterizes these pairs of contraries rests upon an even more fundamental unity. As "beginning and end are together in the circumference of a circle" (fr. 103), so is each pair of opposites contrary only because they share a common ground. This unapparent connection is the basis of apparent difference (fr. 54).

Heraclitean being is a unity of all these pairs of contraries; just as it is not characteristically an aggregate of individual substances, so is it not a collection of disparate particles, or a mechanical system whose lawfulness is ascribed to it or imposed upon it by an adjacent mind, human or divine. Guthrie is correct in observing that the mind/matter division is close to being made, although I think he is mistaken in believing that Heraclitus needs that distinction (*HGP*, 428). Heraclitus anticipates some such distinction sufficiently to specify that all things are one, but not, apparently, enough to include material and immaterial among his pairs of opposites. Another or a later theorist might describe the Logos as immaterial, and might then need to define and explain its relationship to nature. But the Logos is not separate. It is integral in Heraclitus' monism. It is precisely what is common among things, if they are seen as many; it is order in the cosmos when that is seen as one whole.

I take Heraclitus to be an empiricist in so far as he values what is visible and audible. Thus, "things of which there is sight, hearing, knowledge, these I honor most" (fr. 55). One might also adduce as evidence of empiricism such fragments as those which assert that "the sun is new every day" (fr. 6), and that it is the width of a man's foot (fr. 3). The remark that "if all existing things turned to smoke, the nostrils would distinguish them" (fr. 7) seems also to fit in with emphasis on sense awareness. But Heraclitus does not believe that what exists can always be simply located. The seasons cannot be described in terms of simple spatial boundaries, but "the

[1] See Kirk, *HCF*, 72–262 for an extensive analysis of types of opposition. Cf. G. E. R. Lloyd, *Polarity and Analogy* (Cambridge: Cambridge University Press, 1966).

seasons produce all things" (fr. 100).[1] Just as his metaphysical monism is not merely materialism, so also Heraclitus' theory of knowledge is not merely naive empiricism. Search is necessary (fr. 101) because "nature loves to hide" (fr. 123). "The eyes and ears are bad witnesses for men who have barbarous souls" (fr. 107), and "men are deceived in the recognition of visible things" (fr. 56). Reason is present in knowledge no less than in being; reason here consists in the insight which recognizes what is important and identifies the common. Reason does not characterize every belief or every claim to authority; indeed the many lack wisdom (frr. 34, 78, 104, 108), not understanding what is common to all (frr. 114, 51). Familiarity with many opinions does not confer wisdom "otherwise it would have taught Hesiod and Pythagoras, and again Zenophanes and Hecataeus" (fr. 40). Nevertheless, habits of study die hard; they complicate the understanding of Heraclitus no less than they interfere with insight into nature itself.

Aristotle treated Heraclitus' fire as substratum; others have taken Heraclitus' references to the preeminence of fire as symbolic, representing change.[2] Both these interpretations of the status fire of reflect the scholarly habits of thought so much deplored by Heraclitus himself. The notion of substratum is, of course, a philosophical abstraction: while no such thing actually occurs, the concept is useful in discourse among the learned. The notion that fire symbolizes change, representing transformation because it is 'consumed by that which it was nourished by' has basis in Heraclitus' references to kindling and quenching (frr. 30, 43), to smoke (fr. 7), and to the flame lit in the darkness (fr. 26); but it overemphasizes the scholar's experience of domesticated fire: the image seems to derive from the blaze on the hearth and the candleflame in the study.

[1] So much for the notion that Heraclitus does not conceive anything which is not corporeal. See, *e.g.*, Guthrie *HGP*, 428 and Kirk, *HCF*, 53.

[2] Guthrie asserts that Heraclitus' message "outran the resources of contemporary language" so that he resorted to symbols (*HGP*, 439). Zeller thinks Heraclitus speaks symbolically, though he does not think that he distinguishes symbol from thing symbolized. See *A History of Greek Philosophy*, trans. Alleyne, (London, 1881) II, p. 20. B. Snell finds that Heraclitus uses "necessary" metaphors, *The Discovery of the Mind*, trans. Rosenmeyer, (Cambridge, Mass.: Harvard University Press, 1953), p. 220.

It would be more consonant with Heraclitus' regular references to the sun and the stars, and his emphasis on natural processes, to visualize fire not so much as a fuel-fed blaze, but rather as solar and stellar.[1] If fire is primarily sidereal, then its preeminence becomes comprehensible. The crucial fragments are those which refer to measure, most especially "Fire is in exchange for all things and all things for it, as gold is for wares and wares for gold" (fr. 90). Gold is the standard by which wares are measured; fire, here the heavenly bodies, measures earth and sea (Cf. fr. 31), and for that matter growing things, in seasonal and diurnal change. That is to say, the movements of the sun and stars mark out the time for the alternation of day and night, and the changes of the seasons. The same notion is present in fr. 66, although in somewhat garbled form. "Fire, coming upon all things, will judge and sieze them" is stated in terms of the later doctrine of ecpyrosis.[2] Its Heraclitean basis, however, must lie in the notion that fire, that is to say the heavenly bodies, measures the course of nature; it is closely related to fragment 100, which asserts that "The seasons produce all things." The important fragment 30 describes the whole cosmos: "This world order, the same for all, neither any of the gods, nor any man has made, but it was always, and is, and will be an everliving fire kindling in measures and going out in measures." To understand this statement principally in terms of the stipulation that the sun is new every day (fr. 6) is to overemphasize what one might

[1] It is certainly true that from fr. 6 ("The sun is new every day") it follows that the sun is a fuel-fed blaze; the point here is that we cannot by emphasizing that consideration understand why fire has special status in Heraclitus' theory.

In *HCF*, Kirk argues convincingly that the fragments referring to fire emphasize measure and regularity. He explains the special status of fire, however, by taking πῦρ αἰέξωον as αἰθήρ (p. 316). This move seems unwarranted as G. Vlastos shows, "On Heraclitus", *American Journal of Philology*, LXXVI (1955), 362 ff.

[2] K. Reinhardt shows that this statement reflects a Christian interpretation which is based on a Stoic reading of Heraclitus. "Heraklits Lehre von Feuer," *Hermes*, 77 (1942), reprinted in *Vermachtnis der Antike* (Gottingen: Vanderhoeck and Ruprecht, 1966), cited hereafter as 'VA', p. 65. Kirk discusses Reinhardt's argument in *HCF*, 359–60. That this fragment is rooted in something which Heraclitus said seems clear because of Hippolytus' use of φησί, which generally introduces a quotation, though not necessarily. M. Marcovich, *Heraclitus* (Merida: Los Andes University Press, 1967) defends fragment 66, interpreting it as one of the threats with which Heraclitus menaced the Ephesians (p. 436).

call anachronistically the chemical aspect of the process. To take fr. 30 together with 100 and with 94 is to redress that imbalance. According to fr. 94, "The sun will not overstep his measures, otherwise the furies, the ministers of justice, will find him out." These three fragments refer not only to the amounts of earth and water transformed into fire, but also to such cosmic processes as the change of seasons. The position and path of sun or stars across the heavens is regular and dependable. Heat and cold, as well as drought and moisture (fr. 126) are measured by the sun's position, and that in turn has basic limits by which it is bound (fr. 120). The sun's behavior is analogous to that of the river; as fresh waters flow in the river, so also fresh fire blazes in the sun. Both follow regular channels, and may be stronger or weaker according to the season.

Kirk argues that the river statements are to be understood exclusively in terms of the principle of measure. Rejecting the doctrine of flux, he discounts Plato's knowledge of Heraclitus and with it the view that the Ionians were process philosophers. Kirk writes

The preservation of the river's identity and name, in spite of the constant change of its parts, is due to the regularity and balance of that change, just as the preservation of a κόσμος is due to the μέτρα, which govern all meteorological and cosmological change. Plato and all later ancient critics took the river-analogy to apply to changes in every individual thing, and to illustrate the continuity of those changes: actually it illustrates the measure which must inhere in large-scale changes taken as a whole. Heraclitus did not believe, any more than any of his predecessors, that everything was changing all the time, though many things are so changing and everything must eventually change (*HCF*, 366).

I believe that Kirk correctly emphasizes order and measure, but that he underestimates the ancient commentators because he has not genuinely appreciated a characteristic of Ionian monism which Plato did recognize: if distinctions within the one are not real, so that no precise boundaries occur, and if there is change, then flux prevails.

To examine the place of process in Heraclitus' views, it will be well to consider three passages dealing with the river, which are taken as fragments by Diels-Kranz:

Upon those who step in the same river, other and again other waters flow (fr. 12).

One cannot step twice into the same river ... it disperses and gathers, it comes together and flows away, approaches and retreats (fr. 91).

Into the same river we step and do not step. We are and are not (fr. 49a).

Kirk takes fr. 12 to be the basic statement; he interprets it as a claim for the coincidence of "stability (of the whole river) and change (of the waters flowing past a fixed point), rather than continuity of change" (*HCF*, 377). Thus he finds that Plato's interpretation, that "everything is undergoing every motion all the time" (*HCF*, 376; *Theaetetus* 182a), is a caricature of Heracliteanism; Kirk prefers to follow Reinhardt (*VA*, 60) in asserting that "there never was anything approaching a *Flusslehre* in Heraclitus himself" (*HCF*, 370). The defense of his interpretation rests upon the claim that Heraclitus is an empiricist, and we do not in fact observe constant change; it also rests upon the importance of the doctrine of measure, the basis of cosmological structure.

Not all change preserves such an identity, and here a special quality of rivers is relevant; only because the waters flow regularly and replace each other by balanced amounts is the identity [the same river] preserved. This, of course, is precisely the principle of μέτρον which was detected in the cosmological fragments ... in the turnings of fire in fr. 31 the sea is *measured* into the same *proportion* as previously existed, the cosmic fire is kindled *in measures* and extinguished *in measures* in fr. 30. Only by the preservation of these measures, these quantitative material proportions, can the κόσμος or world-order survive at all. That it does persist is demonstrated by our senses; *how* it persists is learned by apprehension of the Logos, a wider aspect of that μέτρον or measure which characterizes every physical change. ... To have said that everything changes, like rivers, would have been for him either an absurdity or a loosely expressed commonplace: what he did say was that natural changes occur in the way that rivers change, i.e. in measures, and thereby maintain in spite of change the unity of the whole κόσμος and the balance of its essential constituents (*HCF*, 377–9).

It is surely true that the river statements are analogous to the cosmological fragments. The difficulty is the supposition that Heraclitus was a materialist. The ascription to Heraclitus of a doctrine of "quantitative material proportions" requiring a "balance of essential constituents" in the cosmos is an unjustified imputation. We cannot say with any sort of faithfulness to Heraclitus that his cosmos has any "essential constitutents," because the cosmos is more truly one than many. The essential character of the cosmos is unity; difference between things is more apparent than real. Although he values the testimony of the senses, Heraclitus is not an empiricist if that label is taken as referring to those who believe that things are just as they appear at first glance. That we do not observe everything changing all the time is not sufficient grounds for rejecting the authenticity of the doctrine of flux. That "quantitative material proportions" are preserved reflects another misapprehension. This phrase suggests that the material – which seems here to be physical bodies, especially earth, fire and water – has a more fundamental importance than the one. The orderliness of change must be granted if we are to follow Heraclitus' instructions; however, priority goes to the one and not to the many. "There is one ordered universe common to all" (fr. 89),[1] he writes; "Listening not to me but to the Logos it is wise to acknowledge that all things are one" (fr. 50). That appearances flow into one another without precise boundaries emerges from fr. 67: "God is day-night, winter-summer, war-peace, surfeit-hunger. But he changes like (fire) which when it mingles with perfumes is named according to each man's pleasure". It is worth noticing that none of the four pairs of opposites cited in this fragment is a body; they might more aptly be described as events or conditions. Nonetheless, Heraclitus takes them as analogues of fire, which suggests that he was interested in the relationship of conditions rather than in a substratum or in individual bodies. Unity takes on various appearances just as the scent of smoke alters. Appearances are shifting; their orderliness is preserved, and the cosmos has a regular structure, by virtue of the sun and the stars, which measure and steer the

[1] Kirk casts doubt upon fr. 89 (*HCF*, 63), which is, however, well defended by Vlastos, pp. 344 ff. Vlastos' valuable article is reprinted (Part I merely summarized, Part II complete, with minor revisions by the author), in *Studies in Presocratic Philosophy*, I, ed. Furley and Allen (London: Routledge and Kegan Paul, 1970).

course of events. The ease with which an Ionian monist can conceive the flow of conditions is clearly expressed by Diogenes of Apollonia, who, at the end of that tradition, defended the old world view

> My view is, in sum, that all existing things are modifications of the same thing, and are the same thing. This is obvious; for if the things now existing in this cosmos – earth, water, air, fire and all the other things which manifestly exist in this cosmos – if one of these was different from another, different, that is, in its own proper nature, and not the same thing changed and altered in many ways, they could in no way mix with one another, nor affect one another for good or ill. Not even a plant could grow from the earth, nor could any animal or anything else come into existence, if everything were not so constituted as to be the same. But in fact all these things are variations of the same thing: they became different at different times and revert to the same thing (fr. 2, transl. Guthrie).

He goes on to say that reason is not separate, but integral:

> without intelligence it would not have been possible for a division to be made in such a way as to keep the measures of all things – winter and summer, night and day, rain, winds, and fair weather; and all else too, if one is willing to reflect, one will find arranged in the best possible way (fr. 3, transl. Guthrie).

Of course Diogenes argued further that being is identifiable as air, a view which cannot be ascribed to Heraclitus. What these fragments do show clearly, however, is that when nature is conceived as a unity in process, one state blends with another by easy transition.

What Plato recognized is that no definite division or precise distinction is correctly applicable to being as the Ionians conceived it. If earth, water, air and fire were "different in their own proper nature" then becoming and change, growth and generation would, as Diogenes points out, have obstacles to overcome. No such obstacles occur for the Ionian monist; consequently his world can appropriately be described as one in which things are always becoming, and in which one stage in process blends with another (*Theaetetus* 152e). Certainly Plato caricatures the position of Heraclitus; it does not follow that he misunderstood altogether what Heraclitus had to say. What Plato does grant is the importance of Heraclitus as a spokesman for the Ionian philosophers. The simultaneity of the one and the many (*Sophist* 242d), the easy transition and the blending of states in continuous flux are genuinely Heraclitean doctrines. Plato does fail to grant or indeed to mention the Heraclitean doctrines of Logos and measure: having distinguished rational entities from perceptible ones and having devoted his attention to the project of making definitions precise, Plato cannot see the cosmos with the eyes of an Ionian, for whom reason and given experience are unified, and for whom nature is one.[1]

[1] I am obviously indebted to the interpretations of Heraclitus' work made by Cherniss and Vlastos, both of whom see Heraclitus as a part of the Ionian tradition, and both of whom defend Plato's insight into Heraclitus' Delphic utterances. My debt to Harold N. Lee is even greater; he has championed Heraclitus for years; his conversation has enabled me to recognize the relation between an indeterminate given unity and the flux. I also wish to thank Louise N. Roberts and R. M. Frazer, Jr., who have suggested a number of improvements in exposition.

KANT'S THEORY OF SYMBOLISM

JOHN D. GLENN, JR.
Newcomb College, Tulane University

Kant's conception of the constructive role of the human mind in experience is widely recognized as one of the historical sources of the various theories of symbolism which have been developed in the twentieth century, such as Ernst Cassirer's "philosophy of symbolic forms." Professor Charles W. Hendel has pointed specifically to the close relation of Cassirer's "symbol" to Kant's "schema."[1] In this light, I think it is interesting to note that Kant himself presented an explicit, but extremely brief, account of the nature of symbolism, and that he distinguished sharply between a schema and a symbol. An examination of Kant's theory of symbolism in the context of his philosophy would seem to be in order.

The Kantian theory of symbolism is presented in a few brief paragraphs of the *Critique of Judgment,* entitled "Of Beauty as the Symbol of Morality."[2] As the title suggests, Kant's chief concern is not to explicate the nature of symbolism, but rather to define the relation between beauty and moral goodness – an important and difficult topic in his aesthetics. But this requires that he first give some account of symbolism in general. He begins with a statement concerning the relation of concepts and intuitions:

In order to demonstrate the reality of our concepts, intuitions are always required. If these are empirical concepts, then the latter are called *examples*. If the former are pure concepts of the understanding, the latter are called *schemata*. If one wishes even to establish the

[1] See Hendel's "Introduction" to Cassirer's *Philosophy of Symbolic Forms*, translated by Ralph Manheim (New Haven: Yale University Press, 1965), I, pp. 12–15.

[2] There are also some general remarks dealing with the nature of symbolism in # 38 of Kant's *Anthropologie*, but the discussion in the *Critique of Judgment* is much more comprehensive.

objective reality of concepts of reason – i.e., the ideas – for the sake of the theoretical knowledge of them, then one desires something impossible, for absolutely no intuition adequate to them can be given.[1]

This paragraph is an admirably concise summary of some of the epistemological doctrine of the *Critique of Pure Reason.* Kant had noted there that, in the case of many empirical concepts, "experience is always available for the proof of their objective reality."[2] But such an "empirical deduction" could not suffice for the pure concepts, or categories, which are employed *a priori.* Kant thought that a "Transcendental Deduction" was necessary to establish their validity, and a "Schematism" to explain their application to objects of experience. He argued that, as empirical concepts find the condition for their applicability to objects in empirical intuition, pure concepts find the condition for their applicability in transcendental schemata, which are *a priori* determinations of pure intuition. In both cases, some sort of intuition is necessary to establish not merely the "objective reality" or "truth" of our concepts, but even their *meaning.* If a concept lacks reference to any object of possible experience, Kant argues, it "has no meaning and is completely lacking in content" (*CR,* A 239/B 298).

From this statement, it would seem to follow not only that experience could never establish the "objective reality" or "truth" of an idea of reason – which Kant defines as "a concept ... transcending the possibility of experience" (*CR,* A320/B377) – but that the ideas lack even *meaning.* This, however, would be inconsistent with the regulative status which Kant claims for certain ideas in the first *Critique,* as well as with his arguments in the second *Critique* that the ideas of God, freedom, and immortality are justified "practical postulates." In the first *Critique,* he deals with this problem by adopting a doctrine of *analogy;* in referring to what transcends possible experience, he says, "the only resource

[1] Kant, *Critique of Judgment,* # 59, p. 254. References to this work are according to the pagination of the third edition, and will be given in parentheses in the text preceded by the abbreviation "*CJ.*" Translations are my own.

[2] Kant, *Critique of Pure Reason,* translated by Norman Kemp Smith (London: MacMillan, 1963), A84/B116. References to this work will be given in parentheses in the text preceded by the abbreviation "*CR.*"

remaining to us is the use of analogy, by which we employ the concepts of experience in order to form some sort of concept of intelligible things" (*CR*, A566/B594). In the *Prolegomena*, he argues, specifically, that analogy enables us to form some sort of concept of God.[1] As we will see, the theory of symbolism presented in the third *Critique* is closely related to this doctrine of analogy.

By asserting, in the paragraph quoted above from the *Critique of Judgment*, that "absolutely no intuition adequate to" an idea of reason is possible, Kant is thus remaining true to the essence of the epistemological position of the third *Critique*. But he does proceed to qualify it significantly by suggesting that there *can* be intuitions which in a way *do* correspond to ideas of reason, that the latter are capable, not of a real exemplification, but of a symbolic presentation in sensible intuition. The sensible presentation, or "*hypotyposis*," of a concept has, he says, two distinct modes:

> It is either *schematic*, where to a concept conceived by the understanding the corresponding intuition *a priori* is given,[2] or *symbolic*, where to a concept which reason alone thinks and to which no sensible intuition can be adequate, an intuition is supplied with regard to which the procedure of the judgment is merely analogous to that which it observes in schematizing – i.e., agrees with it according to the rule of this procedure, not according to the intuition itself, hence merely according to the form of this reflection, not the content (*CJ*, # 59, 255).

In attempting to clarify this matter further, Kant later says that, in symbolic apprehension, "judgment performs a double function," first comprehending an object of sensible intuition, "and then applying the mere rule of reflection on that intuition to a wholly different object, of which the first is only a symbol" (*CJ*, 256).

On this basis, Kant proceeds to distinguish a *symbol* from a *sign*. Criticizing "the new logicians" who contrast the "symbolic" and the "intuitive" modes of representation, he

[1] Kant, *Prolegomena to Any Future Metaphysics*, translated by Lewis White Beck (Indianapolis: Library of Liberal Arts, 1950), # 57–58, pp. 105–106. Kant once uses the term "symbolical" in this context.

[2] Here Kant seems to forget about empirical concepts and their exemplification.

argues that the symbolic is "one species of the intuitive." Intuitive representation "can be divided into the *schematic* and the *symbolic* modes of representation," both of which are real presentations of a concept, and not merely "designations of the concept through accompanying sensible signs that contain nothing belonging to the intuition of the object" (*CJ*, 255). A symbol is in some way analogous to that which it symbolizes, whereas a sign does not have such a relation to what it signifies.[1] Kant provides an example in which one empirical object symbolizes another:

Thus a monarchical state is represented by a living body when it is ruled by internal folk-laws, but by a mere machine (such as a hand-mill) when it is ruled by a single absolute will – but in both cases [it is represented] only *symbolically*. For there is, indeed, no similarity between a despotic state and a hand-mill, but yet there is [such a similarity] in the rules for reflecting about both and their causality. This matter has been little analyzed until now, so it deserves a deeper investigation; but this is not the place to dwell on it (*CJ*, # 59, 256–257).

Clearly this conception of symbolism is closely related to the doctrine of analogy which Kant had stated in the first

[1] I think it is instructive to note that Kant here distinguishes between a symbol and a sign in a different manner than do some of the philosophers whom he has influenced. Cassirer, for example, takes "signs" or "signals" to be essentially stimuli which function in evoking very limited responses, whereas "symbols" bear a wider and more variable meaning. He says that "a signal is part of the physical world of being; a symbol is part of the human world of meaning." (*An Essay on Man* [New Haven: Yale University Press, 1962], pp. 31–36.) Kant's use of the term "symbol" is much narrower than this; for him, even mathematical "symbols," which certainly belong to the "human world of meaning," would be "signs" and not symbols in the proper sense, for they have no intrinsic relation to that which they signify.

Paul Tillich is perhaps somewhat closer to Kant on this point. Where Cassirer seems to make an essentially *functional* distinction between symbols and signs, Tillich attempts to establish an *ontological* distinction. Stating that both symbols and signs point beyond themselves, he argues that the difference between them is "the participation in the symbolized reality which characterizes the symbols, and the non-participation in the 'pointed-to' reality which characterizes a sign." He proceeds to give the example of a flag, which "participates in the power of the king or nation for which it stands and which it symbolizes." (*Theology of Culture* [New York: Oxford University Press, 1964], pp. 54–55.) Tillich is like Kant in asserting that a symbol is somehow intrinsically related to what it symbolizes; however, his unclarified use of the term "participation" and his example of a flag do little to explain his meaning further.

Critique; symbolic representation involves an analogy between the symbol and the symbolized. Nevertheless, I believe that his theory of symbolism does go significantly beyond his earlier-stated doctrine of analogy. For analogy, as Kant had described it in the *Critique of Pure Reason,* involves a relation between two concepts, whereas symbolization involves the relation of an *intuition* to a concept. And it cannot, I think, reasonably be argued that this intuitive element is inessential, that it is *merely* the exemplification of a concept which bears an analogy to the concept symbolized; this would be to neglect the context of Kant's theory of symbolism. For the *Critique of Judgment,* and in particular the "Critique of Aesthetic Judgment," is concerned with *reflecting* judgment, with reflection on an object of intuition which cannot be adequately characterized by any determinate concept. In this context, Kant's theory of symbolism is richer than the doctrine of analogy; it suggests that a symbol is a concrete object of sensible intuition which possesses an overplus of meaning that is not reducible to conceptual meaning or to abstract conceptual analogy.

This is even clearer when Kant's discussion of symbolism in Section 59 of the *Critique of Judgment* is recognized as an extension of his remarks concerning aesthetic ideas and aesthetic attributes in Section 49 of that work. According to Kant, beauty is the expression of "aesthetic ideas," which are representations of the imagination to which no determinate concept is adequate. These are termed "ideas," he explains, because they "at least strive for something lying outside the bounds of experience, and seek to approximate a presentation of concepts of reason (intellectual ideas), which gives the latter the appearance of an objective reality," and also because "no concept can be fully adequate to them." He continues:

> These forms which do not comprise the presentation of a certain concept itself, but which – as near-representations [*Nebenvorstellungen*] of the imagination – express only the consequences connected with it, and its connection with other [concepts], can be called (aesthetic) *attributes* of an object whose concept, as a rational idea, cannot be adequately presented. Thus Jupiter's eagle with lightning in its claws is an attribute of the king of heaven, and the peacock of the splendid heavenly queen. They do not, like *logical attributes,* set forth what is

contained in our concept of the sublimity and majesty of creation, but something else, which gives occasion to the imagination to diffuse itself over a mass of connected representations that let more be thought than can be expressed in a concept determined by words ... (*CJ*, # 49, 193–195).

Because such "forms" express more than can be adequately summed up in any concept, Kant's theory of symbolism seems to be a significant modification of his earlier-stated doctrine of (conceptual) analogy. His realization that more can be expressed in concrete symbolic form than can be explicated conceptually is highly significant in light of the fact that Kant's temperament, if not his doctrine, had always been rationalistic. Moreover, this insight seems to contain the possibilities for some rather fundamental modifications in the critical philosophy. The outlines of some of these possibilities begin to emerge in the further discussion of symbolism in Section 59 of the *Critique;*

Our language is full of such indirect presentations according to an analogy, in which the expression does not contain the proper schema for the concept, but merely a symbol for reflection. Such are the words "ground" (support, basis), "depend" (to be held from above), "flow from" (instead of "follow"), "substance" (as Locke expresses it: the support of accidents), and countless other not schematic but symbolic hypotyposes and expressions for concepts, not by means of a direct intuition, but only according to an analogy with it ... (*CJ*, # 59, 257).

Kant proceeds to say that, if the practical determination of what the idea of God is to be for us can be called "knowledge," "then all of our knowledge of God is merely symbolic." He adds that, if these symbols are taken literally, as schemata, anthropomorphism is the result, while he who "omits everything intuitive" falls into deism, "through which nothing at all is known, even for a practical purpose" (*CJ*, # 59, 257–258).

It can hardly be mere accident, I think, that the examples which Kant cites here suggest what are, according to his philosophy, among the most basic concepts of the human mind – substance and causality. Perhaps he did not intend or was not fully aware of some of the implications of his remarks, but they seem at least to hint at the possibility that these categories are not the fixed products of an abstract understanding, but rather have their origin in certain "root metaphors," certain symbolic apprehensions, and that *schemata* may be merely pale reflections of what were originally *symbols*.

Kant's suggestion that "all of our knowledge of God is merely symbolic" does not, of course, mean that the *truth* of assertions about God can be established through some sort of symbolic insight; he clearly continues to think that this remains in the realm of a morally-based faith. But he is claiming that the *meaning* of such assertions is basically symbolic. Kant may mean only that the representation of God is by means of conceptual analogy,[1] which would mark no change from the position which he had developed earlier. But what he says *may* also reflect the recognition that abstract analogical conception of God requires enrichment by concrete religious symbolism if it is to have real significance for human life.[2]

The overall tendency of Kant's discussion of symbolism[3] is, I think, to suggest that man's higher mental faculties, understanding and reason, are much more closely intertwined with imagination and sense than he had earlier thought, that the basic concepts (the categories and the ideas of reason) which the human mind employs can only come to be conceived because they are capable of being first presented in concrete symbolic form. From the perspective of the twentieth century, of course, these ideas are not startlingly new. But

[1] Kant had used the term "symbolical" as a mere synonym for "analogical" in discussing the analogical conception of God in the *Prolegomena*. (See above, page 15, footnote 1.) This is the more probable meaning here.

[2] Cassirer has noted that Kant's rational religion is "only the ideal shape, only the shadow, of what a genuine and concrete religious life is." (*Essay on Man*, p. 25.)

[3] I will not here discuss in detail Kant's application of his theory of symbolism in his claim that beauty is the symbol of morality; that topic deserves separate treatment. I should say, however, that I am convinced that this claim does not conflict with the notion, expressed at other points in the *Critique*, that beautiful objects can symbolize various rational ideas. The symbolization in these two contexts is on different levels. In the latter, Kant is suggesting that beautiful objects can be symbolic presentations of different rational ideas. In the former, he is fundamentally pointing to an analogy between the aesthetically judging subject and the morally willing subject. These two perspectives are, I think, completely compatible.

this is because of the development of the philosophy of symbolism by thinkers who directly or indirectly owe much to Kant. This is especially true of Ernst Cassirer, who has made symbolism the key to a comprehensive interpretation of human culture.

Yet it must be noted that there is one aspect of Kant's theory of symbolism which is lacking, or at least unstressed, in Cassirer's. This is the claim that symbols enable us to *think*, although not to know, what transcends our experience. This is perhaps the most basic significance of Kant's distinction between a schema and a symbol. Whereas a *schema* functions in the *constitution* of experience, a *symbol* is an object of experience which somehow refers *beyond* the limits of experience. The fact that Cassirer does not make such a distinction, that his "symbol" is, despite its greater concreteness, closely related to Kant's "schema," indicates that he takes symbols to be constitutive of experience, and does not acknowledge their capacity to refer beyond the experienced.

The difference between the two philosophers may be illustrated by referring to a passage in which Cassirer urges that we "accept in all seriousness what Kant calls his 'Copernican revolution,' that "instead of measuring the content, meaning, and truth of intellectual [symbolic] forms by something extraneous which is supposed to be reproduced in them, we must find in these forms themselves the measure and criterion for their truth and intrinsic meaning."[1] If this is what it means to accept the Copernican revolution "in all seriousness," it must be said that Kant himself never completely did so! While granting to the schema, as the medium through which a category is applied to experience, status as a locus of meaning and truth, he restricted its significance to the realm of possible experience. He held to the idea – which the neo-Kantians rejected – of the thing-in-itself, of a dimension of reality which is in no way constituted by the human mind, and claimed for the symbol the power to give *meaning* to our references to this transcendent realm, although the *truth* of such references remained a matter of moral faith.

[1] Ernst Cassirer, *Language and Myth*, translated by Suzanne K. Langer (New York: Dover, 1953), p. 8.

It would be too much to expect to decide the issue between Kant and Cassirer here. But it seems worthwhile to reflect on the significance of the difference between them. Cassirer's theory of symbolism represents a development of Kantian thought in an idealistic and anthropocentric direction.[1] It expresses an optimism about the ability of man to endow the universe with meaning through the use of his own mental powers. Kant's philosophy, however, respects what he considered to be the limits of human knowledge.[2] It expresses the conviction that, despite human freedom and creativity, human existence is ultimately in relation to a transcendent ground and in need of a symbolically-expressed practical faith.[3] Thus Kant's theory of symbolism does not so much look forward to the idealistic development of his theory of knowledge as it recalls the medieval doctrine of analogy and the conviction that human existence requires an orientation toward the transcendent.

[1] Here Cassirer is close to Hegel. See Hendel's above-cited "Introduction" for several remarks on the relation of Cassirer to Hegel.

[2] For an interpretation of Kant's philosophy as a philosophy of human finitude, see Heidegger's *Kant and the Problem of Metaphysics*, translated by James S. Churchill (Bloomington, Ind.: Indiana University Press, 1962).

[3] The concept of a transcendent ground appears in all three *Critiques*, although in somewhat different forms.

DUALISMS IN WILLIAM JAMES'S *PRINCIPLES OF PSYCHOLOGY*

ANDREW J. RECK
Tulane University

I. *Introduction*

Twentieth century American philosophy has been characterized by what one of our best philosophical historians, Max Fisch, has called "the damnation of Descartes."[1] A conspicuous facet of American anti-Cartesianism has been the rejection of dualisms, and in this regard American thought has been part of an international "revolt against dualism."[2] Harold N. Lee, whose own thought is a creative outgrowth of American pragmatism and naturalism, has joined arms with those who would make the "revolt" into a successful "revolution." Lee has acknowledged Peirce, Mead, and C. I. Lewis to be his predecessors who have influenced him in his efforts to overcome dualisms.[3]

Yet few American thinkers can match William James in the weight of their contributions to the rejection of dualisms. James's 1903 essay, "Does 'Consciousness' Exist?," seemed to a generation of American philosopher-psychologists to put dualisms to rest.[4] And his major work, *The Principles of Psychology* (1890), was opposed at once to (1) the associationism of the empirical school with its atomic ideas, and (2) the

[1] Max Fisch, "General Introduction," *Classic American Philosophers* (New York: Appleton-Century-Crofts, 1951), pp. 19 ff.

[2] A. O. Lovejoy, *The Revolt Against Dualism* (La Salle, Ill.: Open Court Publishing Company, 1930).

[3] See Harold N. Lee, *Percepts, Concepts, and Theoretic Knowledge*, to be published by Memphis State University Press.

[4] First published in *The Journal of Philosophy* (1903); reprinted in *Essays in Radical Empiricism*, ed. by Ralph Barton Perry (New York: Longmans, Green, 1912). For a critical response by one thinker who did not succumb to James's theory, see A. O. Lovejoy's essay, "James's 'Does Consciousness Exist?'," published in *The Thirteen Pragmatisms and Other Essays* (Baltimore: The Johns Hopkins Press, 1963), pp. 112–132.

faculty theories of the Kantian and post-Kantian transcendentalists as well as of the Cartesian spiritualists. More positively stated, James's achievement was to work out in psychology the implications of Darwin's theory of biological evolution, as others, including Harold N. Lee, have sought to do in epistemology. According to the evolutionary view of life and mentality, the fundamental situation is that of an organism struggling with its environment in order to survive. As James put the matter in his essay, "Rationality, Activity and Faith" (1882): "The theory of Evolution is beginning to do very good service by its reduction of all mentality to the type of reflex action. Cognition, in this view, is but a fleeting moment, a cross-section at a certain point of what in its totality is a motor phenomenon."[1]

Following in the footsteps of Herbert Spencer, whose *Psychology* he had used as a text in his own courses, James accepted the evolutionary account of the naturalistic status of mind (or consciousness) which guages the different degrees and kinds of mentality by the minded organisms' adjustments to the environment. But James rejected Spencer's theory insofar as it construed mind in wholly mechanistic terms and reduced it to a passive faculty adjusting to a fixed environment. Although James reinforced the naturalistic conception of the dependence of mind upon cerebral conditions, he nevertheless stressed that mind is teleological. In the first chapter of *The Principles of Psychology* he defined mind (or consciousness) in terms of purposiveness. The mark and criterion of mind is, he declared, "*the pursuance of future ends and the choice of means for their attainment*..." (*P*, I, 8). Hence mind is a function of the organism which enables it not only to adjust to its environment but also to modify its environment in line with its own ends.

Now James's *Principles* marks the opening of a new era in American psychology. It initiated the movement called "functionalism," which in turn gave way to "behaviorism." It seemed to deal Cartesian dualism a lethal blow. For James

[1] Quoted in William James, *The Principles of Psychology* (New York: Holt, 1890), II, p. 313. Hereafter all references to the *Principles* will be made in parentheses in the text, symbolized by "*P*" followed by volume and page numbers.

sketched a general theory of mind which, on the one hand, made mind integral to the living organism, and which, on the other hand, prepared the way for the elimination of mentalistic concepts in psychological explanations of behavior. In sum, James not only offered a new theory of mind – functionalism; he also set out to establish psychology as a natural science.

James was explicit about his intentions in the Preface to the *Principles*. In seeking to establish psychology as a natural science, he recognized that as such it occupied a restricted context within the broader domain of philosophy.[1] Further, he was aware that psychology as a natural science rests upon assumptions, and he took care to state these assumptions: "Psychology ... assumes as its data (1) *thoughts and feelings*, and (2) *a physical world* in time and space with which they coexist and which (3) *they know* ... This book, assuming that thoughts and feelings exist and are vehicles of knowledge, thereupon contends that psychology when she has ascertained the empirical correlation of the various sorts of thought and feeling with the definite conditions of the brain, can go no farther ..." (*P*, I, v). In a very illuminating recent work, *The Pragmatic Movement in American Philosophy*, Charles Morris has referred to these assumptions of James's *Principles* as "methodological dualism."[2] The topic of the present paper is an examination of James's dualistic assumptions, to correct any hasty inference that James was always anti-dualistic and to reveal, by implication, just how entrenched dualism was, even in the mind of one of its leading adversaries, when it was first challenged. Moreover, Morris's rubric "methodological dualism" obscures the complexity and depth of James's dualism. For however much James may have ultimately contributed to the overcoming of dualism, at the time he wrote the *Principles* he was a dualist. And he was a dualist in that twofold sense so ably distinguished by A. O. Lovejoy: a psycho-physical dualist and an epistemological dualist.

[1] I have treated this point in my paper, "The Philosophical Psychology of William James," *Southern Journal of Philosophy*, Vol. 9 (1971), 293–312.

[2] Charles Morris, *The Pragmatic Movement in American Philosophy* (New York: George Braziller, 1970), p. 123.

II. Psycho-physical Dualism

Throughout *The Principles* James adhered to psycho-physical dualism. His stipulation that psychology be a natural science explicitly entailed that the psychologist as such has "nothing to do" with "Mind unattached to a particular body" (*P*, I, 183). Psycho-physical dualism is the doctrine that there are two principles, mind (or consciousness) and body (especially the brain), and that psychology as a natural science studies both and their relations to each other. Psycho-physical dualism need not hold that these principles be substances, and in James's psychology they are, in fact, processes. However, psycho-physical dualism has considerable latitude in regard to the kind of relations subsisting between its two principles. But James in general terms maintained that mind (or consciousness) is related to its body dynamically – i.e., causally; although he took pains to show how various the lines of causation are for different mental and physical activities. At any rate, psycho-physical dualism pervades James's *Principles*.

To a considerable extent this dualism is buttressed by the methods which James employed in psychology. In Chapter VII of *The Principles* he listed three methods: the introspective, the experimental, and the comparative. The introspective method is the method which, said James, "... we have to rely on first and foremost and always" (*P*, I, 185). This method consists in "... the looking into our own minds and reporting what we there discover" (*P*, I, 185). By means of introspection we discover "states of consciousness." As James stated: "So far as I know, the existence of such states has never been doubted by any critic, however sceptical in other respects he may have been ... *I regard this belief as the most fundamental of all the postulates of Psychology*, and shall discard all curious inquiries about its certainty as too metaphysical for the scope of this book" (*P*, I, 185). Yet James was cognizant of the inadequacies of the introspective method, and he looked to the new experimental method in psychology to remedy its defects. To be precise, he referred to experimentation in the plural, speaking of experimental methods which begin with introspective data but seek to eliminate

... their uncertainty by operating on a larger scale and

taking statistical means" (*P*, I, 192). As regards the comparative method, it simply employs both methods but extends their application to the entire animal kingdom, in order to ascertain "... the origin of these features [the features of normal consciousness revealed in introspection], or their dependence upon one another," by tracing this origin through "... all its possible variations of type and combination" (*P*, I, 194). Now it should be clear from James's account of these methods that he was committed to psycho-physical dualism. Crucial here was his recognition of the validity of the experimental methods, although he was admittedly critical of the results so far attained. Nevertheless, he discerned that they heralded a new era in the history of psychology, marking its establishment as a natural science. At the same time he understood experimentation in psychology to be psycho-physical. Starting from mental data gained through introspection, the experimental method correlated these data with physical processes and events in the body, particularly in the brain and the nervous system.

James examined the problems of psycho-physical dualism in Chapters II through VI of the *Principles*, touching many delicate metaphysical issues. He regarded the dualistic problem of relating body and mind, or brain and consciousness, to be "... the ultimate of ultimate problems" (*P*, I, 177). A brief review of the results of James's discussion is in order here.

The topic of Chapter II is "the functions of the brain." James reviewed the experimental data which showed correlations between parts and processes of the brain and conscious activities or states, such as seeing, hearing, smelling, thinking, and maintained the general thesis, now widely accepted, that mental events are correlated (if not reduced to) brain events. James's particular version of the thesis is non-reductionist and hence emphatically psycho-physical.

In Chapter III James considered "some general conditions of brain-activity." In passing, James examined and rejected various attemps to reduce consciousness to materialistic elements, such as phosphorous. A passage from James is worthy of quotation here:

'The brain secretes thought, as the kidneys secrete urine, or as the liver secretes bile,' are phrases which one sometimes hears. The lame analogy need hardly be pointed out. The materials which the brain *pours into the blood* (cholesterin, creatin, xanthin, or whatever they may be) are the analogues of the urine and the bile, being in fact real material excreta. As far as these matters go, the brain is a ductless gland. But we know of nothing connected with liver – and kidney – activity which can be in the remotest degree compared with the stream of thought that accompanies the brain's material secretions (*P*, I, 102–103).

In Chapter IV James discussed habit. This chapter is redolent with suggestions of a metaphysics of evolutionary will, a theme which recurs in the *Principles* and which, though worthy of investigation, is beyond the compass of the present paper. Now as James employed the term "habit" he meant more or less stable patterns of behavior, including instinctive behavior in animals. Habits are useful in purposive behavior. They simplify the action required to attain an end, mainly by diminishing the amount of conscious attention paid to an action. They also increase accuracy and reduce fatigue. James's explanation of habit in terms of paths grooved into the nervous system is as much physical as physiological, but it is physical, too, since it seems to resort to will or intelligence as the agency which initiates habit – i.e., first makes the paths.

In Chapter V James examined the automaton theory of mind. This theory construes all organic behavior as reflex action. It traces an automatic sequence of occurrences from the stimulation of the sense organs to the discharge of the last brain-vibration into the motor tracts. Thus it denies consciousness any causal efficacy, regarding it to be only "an 'epiphenomenon,' an inert spectator, a sort of 'foam, aura, or melody' as Mr. Hodgson says, whose opposition or whose furtherance would be alike powerless over the occurrences themselves" (*P*, I, 129). Allowing causation from physical brain state to mental state and from physical state to physical state, the automaton theory rules out causation from mental state and even from mental state to mental state. Now James was clear in his repudiation of the automaton theory. As he said, "... to urge the automaton theory upon us, as it is now urged, on purely *a priori* and *quasi*-metaphysical grounds, is an *unwarrantable impertinence in the present state of psychology*" (*P*, I, 138).

James's rejection of the automaton theory stemmed not only from his conviction that its proponents were mired in an unscientific metaphysics, but also from his adherence to a theory of mind which, in agreement with the common sense acceptance of both physical and mental causes, views mind as causally efficacious. James, moreover, adapted this common sense view to the theory of evolution. Since the higher animals possess more complex and intense consciousness than the lower animals, the theory of evolution induces us to conclude that consciousness is "... an organ, superadded to other organs which maintain the animal in the struggle for existence; and the presumption of course is that it helps him in some way efficacious and influencing the course of his bodily history" (*P*, I, 138).

How, then, does consciousness help the organism? It does so, according to James, by working on the brain much as a regulator works on a machine. James wrote:

> ... consciousness is at all times primarily *a selecting agency*. Whether we take it in the lowest sphere of sense, or in the highest of intellection, we find it always doing one thing, choosing one out of several of the materials so presented to its notice, emphasizing and accentuating that and suppressing as far as possible all the rest. The item emphasized is always in close connection with some *interest* felt by consciousness to be paramount at the time (*P*, I, 139).

Further, the function of consciousness for an organism is roughly proportional to the complexity of its brain and nervous system. Organisms with complex brains and nervous systems are less stable but more flexible than lower types of organisms. Hence it brings "... a more or less constant pressure to bear in favor of *those* of its [the organism's] performances which make for the most permanent interests of the brain's owner; it would mean a constant inhibition of the tendencies to stray aside" (*P*, I, 139).

In Chapter VI James took up "the Mind-Stuff theory." This theory grew out of associationism in psychology and evolution in biology. Advocated by Herbert Spencer, among others, it contended that "... *our mental states are compounds* [of primordial mental atoms or bits of mind dust], expressed in its most radical form" (*P*, I, 145). James's overriding objection to the theory is that it failed to account for the essential features of all known combinations. As he said, "All the 'combinations' which we actually know are EFFECTS, wrought by the units said to be 'combined,' UPON SOME ENTITY OTHER THAN THEMSELVES" (*P*, I, 158). One consequence of conceiving consciousness to be an integral unity rather than a compound, James noted, is to complicate the task of explaining the relation between brain and consciousness as a unity at a given moment is correlated. Rather the brain is, in James's words, "... a multitude of physical facts. 'Entire brain' is nothing but our name for the way in which a million of molecules arranged in certain positions may affect our sense" (*P*, I, 178).

Nor did James accept the material-monad theory to account for the relation of mind and body, although he conceded its superiority to the mind-dust theory in stressing the unity of consciousness. For according to the monad theory, the brain is a compound of cells, each with its own consciousness, and it contains a central cell upon which all the other cells act and which in turn acts on them, this central cell having a unified consciousness which integrates the consciousness of all the other cells. Nevertheless, James repudiated this theory on scientific and philosophical grounds. Scientifically, no central cell or group of cells can be found. Philosophically, if one could be found, it would in principle be analyzed into simpler units, so that in the final analysis we would find as "... the elementary and irreducible psycho-physical couple, not the cell and its consciousness, but the primordial and eternal atom and its consciousness" (*P*, I, 180).

Thus James reluctantly had to admit the term "soul" into his text. He remarked that no one should despise any great traditional objects of belief since "... there is always a great drift of reasons, positive and negative, towing us in their direction" (*P*, I, 181). Indeed, he acknowledged "the logical respectability of the spiritualistic position ... The soul would be thus a medium upon which ... the manifold brain processes *combine their effects*" (*P*, I, 181). As James said, "I confess, therefore, that to posit a soul influenced in some mysterious way by the brain-states, and responding to them by conscious affections of its own, seems to me the line of least logical resistance, so far as we have yet attained" (*P*, I, 181).

Yet James did not embrace the soul-theory. Scientific psychology does not need it, but can make do with an empirical parallelism between mental states discovered introspectively and physical states amenable to observation and experimentation. Furthermore, in Chapter X, where James dealt with "the consciousness of self," he was clear in regard to the physical inadequacy of the theory of the soul. As James argued, the theory of the soul is tied up with the doctrine of substance. Consequently it is difficult to understand. Worse still, it introduces an element which it claims is more than the flux of consciousness and of physical states, but which, as James asserted, "... *explains* nothing; and when we are once trying metaphysical explanations we are foolish not to go as far as we can. For my own part I confess that the moment I become metaphysical and try to define the more, I find the notion of some sort of an *anima mundi* thinking in all of us to be a more promising hypothesis, in spite of all its difficulties, than that of a lot of absolutely individual souls" (*P*, I, 346).

James retreated from elaborating the metaphysics which emerges intermittently in the *Principles*. Metaphysics, he argued, had no place in psychology as a natural science. When he ventured into the construction of a theory of the self – i.e., of the individual mind or consciousness, he sought to reject all metaphysical theories. Instead of propounding a metaphysical theory of the self as a non-phenomenal entity, then, he proceeded to elucidate in introspective terms the sense each of us has of his own personal identity. This sense, he continued, "... *is exactly like any one of our other perceptions of sameness among phenomena. It is a conclusion grounded either on the resemblance in a fundamental respect, or on the continuity before the mind, of the phenomena compared*" (*P*, I, 335). Accordingly, the "I" which binds together our perceptions and feelings is the present passing thought. As

James said: "The passing Thought ... seems to be the Thinker; and though there *may* be another non-phenomenal Thinker behind that, so far we do not seem to need him to express the facts" (*P*, I, 342). James described the relationship of the passing Thought to its predecessors as a process of appropriation and of dying away. The present passing Thought appropriates its predecessors and passes away. As James said, "Each Thought is thus born an owner, and dies owned transmitting whatever it realized as its Self to its own proprietor ... Who owns the last self owns the self before the last, for what possesses the possessor possesses the possessed" (*P*, I, 339). Further, the passing Thought, since it cannot appropriate itself, cannot be its own object. "It appropriates *to* itself, it is the actual focus of accretion, the hook from which the chain of past selves dangles, planted firmly in the Present, which alone passes for real, and thus keeping the chain from being a purely ideal thing" (*P*, I, 340–341).

At this juncture James returned to the body to stabilize his theory of the consciousness of self. Having contended that the Thinker as the passing Thought immediately perishes upon its birth, he observed that "... nothing can be known about it till it be dead and gone. Its appropriations are therefore less to *itself* than to the most intimately felt *part of its present Object, the body, and the central adjustments*, which accompany the act of thinking, in the head. *These are the real nucleus of our personal identity* ..." (*P*, I, 341).

III. Epistemological Dualism

An unquestioned epistemological dualism dominates the *Principles*. James himself has defined epistemological dualism accurately. In Chapter VIII, on "the relations of minds to other things," James asserted: "*The psychologist's attitude toward cognition ... is a throughgoing dualism.* It supposes two elements, mind knowing and thing known, and treats them as irreducible" (*P*, I, 218).

In Chapter VII James had pointed out that the psychologist as natural scientist studies minds as "... *objects*, in a world of other objects" (*P*, I, 183). When, for example, the psychologist studies the perception of color, he distinguishes perception from color and treats both as objects of his study. James, moreover, drew a diagram to exhibit the epistemological "assumptions of psychology" as a science. Each assumption marks off a set of what he called "the irreducible data of psychology." These assumptions are: (1) the psychologist, (2) the thought studied, (3) the thought's object, and (4) the psychologist's reality. Now these four assumptions reveal two instances of epistemological dualism. The first instance, which typifies the form of epistemological dualism in general, occurs between assumption (2), the thought, and assumption (3), the thought's object. The second instance, which specifies epistemological dualism as it pertains to psychology, occurs between assumption (1), the psychologist, and the psychologist's total object, which embraces assumption (2), the thought studied; assumption (3), the thought's object; and assumption (4), the psychologist's reality.

Failure on the part of the psychologist to understand and to keep critically before him the significance of the epistemologically dualistic character of all knowledge, including his own, has produced, according to James, one of the great snares of psychology – namely, "the psychologist's fallacy." This fallacy arises from "... the *confusion of his* [the psychologist's] *own standpoint with that of the mental fact* about which he is making his report" (*P*, I, 196). In other words, there is a confusion of the specific dualism, which consists in the cognitive relation of assumption (1) to assumptions (2), (3), and (4), with the generic dualism, which consists in the cognitive relation of assumption (2) to assumption (3). For James many crucial problems in philosophy, such as the problem of universals, have been compounded by virtue of this confusion. "Another variety of the psychologist's fallacy," continued James, "is the assumption that the mental state studied must be conscious of itself as the psychologist is conscious of it" (*P*, I, 197). In other words, the terms of the generic dualism are themselves confused so that the cognitional value of the psychologist's thought is wrongly imputed to part of his object – namely, the idea that he is studying. More simply put, the psychologist's own thought is confused with his object. Throughout the *Principles* James pointed repeatedly to this fallacy as the source of grave errors.

Whereas the psychologist, on James's account, may employ epistemological dualism critically in the avoidance of fallacies, he just simply accepts this dualism without attempting to explain it. James wrote:

...the *relation of knowing* is the most mysterious thing in the world. If we ask how one thing *can* know another we are led into the heart of *Erkenntnisstheorie* and metaphysics. The psychologist, for his part, does not consider the matter so curiously as this. Finding a world before him which he cannot but believe that *he* knows, and setting himself to study his own past thoughts, or someone else's thoughts of what he believes to be that same world; he cannot but conclude that those other thoughts know it after their fashion even as he knows it after his. Knowledge becomes for him an ultimate relation that must be admitted, whether it be explained or not, just like difference or resemblance, which no one seeks to explain (*P*, I, 216).

Nevertheless, the psychologist is able to solve one of the gravest problems of dualism, at least so far as it concerns his discipline. The problem is: How can a mind know that it knows its object, since there is no standpoint outside the mind to compare its knowledge with its object? Conceding that this problem is insoluble if Mind is Absolute, James argued that, since the scientific psychologist studies finite minds only, he "... himself can go bail for the independent reality of the objects of which they think. He knows these to exist outside as well as inside the minds in question; he thus knows whether the minds think and *know*, or only think; and though his knowledge is of course that of a fallible mortal, there is nothing in the conditions that should make it more likely to wrong (*sic*) in this case than in any other" (*P*, I, 217).

Naturally the psychologist is called upon to decide whether a state of mind he is studying is or is not knowledge. Hence he must employ some tests for knowledge, and these tests, according to James, are

...the tests we all practically use. If the state of mind *resembles* his [the psychologist's] own idea of a certain reality; or if without resembling his idea of it, it seems to imply that reality and refer to it by operating upon it through the bodily organs; or even if it resembles and operates on some other reality that implies, and leads up to, and terminates in, the first one, – in either or all these cases the psychologist admits that the state of mind takes cognizance, directly or remotely, distinctly or vaguely, truly or falsely, of the reality's nature and position in the world. If, on the other hand, the mental state under examination neither resembles nor operates on any realities known to the psychologist, he calls it a subjective state pure and simple, possessed of no cognitive worth. If, again, it resembles a reality or set of realities as he knows them, but altogether fails to operate on them or modify their course by producing bodily motions which the psychologist sees, then the psychologist, like all of us, may be in doubt (*P*, I, 217).

James considered knowledge to be a complex affair. First the thing to be known must causally affect the brain. Then the mind, which is dynamically related to the brain, constitutes knowledge "... by a new construction that occurs altogether *in* the mind" (*P*, I, 219). Thus knowledge is not the mere effect in a line of causation triggered by the thing and transmitted through the sense organs, the nervous system, and the brain. On the contrary, knowledge is an achievement of the mind. As James said, "... even in mere sense-impression the duplication of the object by an inner construction must take place" (*P*, I, 219). What guarantee is there that knowledge, created by the Subject, agrees with its Object? James's response to this critical question is straightforwardly realistic for psychology as a natural science: "The dualism of Object and Subject and their pre-established harmony are what the psychologist as such must assume, whatever ulterior monistic philosophy he may, as an individual who has the right also to be a metaphysician, have in reserve" (*P*, I, 220).

As a psychologist James was not required to explain the metaphysics of the epistemological dualism which he as a scientist assumed uncritically. He was, however, obliged to describe the cognitive state as it occurs in consciousness. In Chapter IX on "the stream of thought" he noted as the fourth characteristic of consciousness the fact that it appears to deal with independent objects. He wrote: "*Human thought appears to deal with objects independent of itself; that is, it is cognitive, or possesses the function of knowing*" (*P*, I, 271).

Turning from the metaphysical question whether or not the objects of thought exist independently, James offered a psychological explanation as to "... why we all believe that the objects of our thoughts have a duplicate existence outside the mind ..." (*P*, I, 271). The reason why we believe so is "... that there are *many* human thoughts, each with the *same* objects, as we cannot help supposing" (*P*, I, 271–272). The basis of our belief, then, is "... *sameness* in a multiplicity of objective appearances" (*P*, I, 272).

James contended that his view that consciousness deals with, intends, or knows independent objects is nothing but "... the *psychological* point of view, the relatively uncritical non-idealistic point of view of all natural science, beyond which this book [*The Principles of Psychology*] cannot go" (P, I, 272). Yet James's line of argument was not naive, much of it anticipating later phenomenological conceptions of the intentionality of consciousness. He asserted that consciousness of objects is primitive, coming before consciousness of itself, and, indeed, he maintained that in certain chloroformed states we have a "... sense of objectivity and lapse of subjectivity" (*P*, I, 273). He ruled out, in particular, the Kantian theory that self-consciousness is essential to consciousness, citing this theory as an instance of the psychologist's fallacy. James stressed that "... *thought may, but need not, in knowing, discriminate between its object and itself*" (*P*, I, 275).

At this juncture James sought to elucidate "... *the proper use of the term Object in Psychology*" (*P*, I, 275). He distinguished the object of thought from its topic. He illustrated his distinction by consideration of the thought expressed in the sentence "Columbus discovered America in 1492." Whereas the topic of this thought may be Columbus or America or the discovery of America,

> ...the *Object*... is really its entire content or deliverance, neither more nor less... The object... in the sentence, for example, is strictly speaking neither Columbus, nor America, nor its discovery. It is nothing short of the entire sentence, 'Columbus-discovered-America-in-1492.' And if we wish to speak of it substantively, we must make a substantive of it by writing it out thus with hyphens between all its words. Nothing but this can possibly name its delicate idiosyncrasy. And if we wish to *feel* that idiosyncrasy we must reproduce the thought

as it was uttered, with every word fringed and the whole sentence bathed in that original halo of obscure relations, which, like an horizon, then spread about its meaning (*P*, I, 275–276).

Hence human thought, according to James, points to – intends – objects beyond itself. Every cognitive state of consciousness primitively posits an Object. Yet, citing Brentano, James distinguished conception from belief. In the case of conception, the object is merely apprehended by the mind. In the case of belief, it is also "... held to have reality. Belief is thus the mental state or function of cognizing reality" (*P*, II, 286). Further, belief is for James the natural condition of the human mind. So long as any object is uncontradicted, it is believed and posited as absolute reality. James, however, listed seven subuniverses of reality, beginning with the world of sense or of physical things and ending with the worlds of sheer madness and vagary. Every object of thought belongs to one or other of the subuniverses. The question: What is real? is, for James, resolved by the choice the individual makes according to his interests. James wrote:

The mere fact of appearing as an object at all is not enough to constitute reality. That may be metaphysical reality, reality for God; but what we need is practical reality, reality for ourselves; and, to have that, an object must not only appear, but it must appear both *interesting* and *important*. The worlds whose objects are neither interesting nor important we treat simply negatively, we brand them as *un*real.

In the relative sense, then, the sense in which we contrast reality with simple *un*reality, and in which one thing is said to have *more* reality than another, and to be more believed, *reality means simply relation to our emotional and active life*. This is the only sense which the word ever has in the mouths of practical men. *In this sense, whatever excites and stimulates our interest is real*... (*P*, II, 295).

Of course James admitted what he called "... the prerogative position of sensations in regard to our belief" (*P*, II, 305). Although every thought has an object, James emphasized that belief consists in the emotional reaction of the entire man toward this object, whatever it may be. While knowledge is dualistic, relating subject and object, the interests of the subject are central in establishing the relation and in imbuing the object with reality. In this sense, belief is one with will. As James said: "*Will and Belief, in short, meaning a certain relation between objects and the Self, are two names for one and the same PSYCHOLOGICAL phenomenon*" (*P*, II, 321).

IV. Summary Conclusions

Dualisms permeate James's *Principles of Psychology*.

Psycho-physical dualism is implicated in James's conception of the valid methods of psychology. It is admitted at every point in the employment of these methods for the description and explanation of conscious states and physical motions of complex organisms. It is invoked in the functionalist theory of mind James proposed. It is even resident in his theory of the consciousness of the self. It is not correct to say, as John Wild does, that James's psycho-physical dualism in the *Principles* is "vague" and that "... there is abundant evidence to show that he [James] was not satisfied with ..." it.[1]

Epistemological dualism James assumed without questioning. Failure on the part of the psychologist to note the terms of this dualism and to keep them distinct is, according to James, the source of "the psychologist's fallacy." Consciousness itself reveals, upon analysis, the intentionality of its states to objects beyond itself. At the same time, these objects are shown to be practically real so far as they relate to the individual knower's practical interests.

James, then, did not reject dualisms, at least in *The Principles of Psychology*. But it should be clear that these dualisms are situated for James in a broader philosophical context, one involving a metaphysics of evolutionary will. The clarification of this metaphysics, let us repeat, is beyond the compass of the present paper, as indeed James, despite numerous allusions to it, deemed all metaphysics to be out of place in psychology as a natural science.

[1] John Wild, *The Radical Empiricism of William James* (New York: Doubleday, 1969), p. 359. Note that Wild readily accepts James's epistemological dualism, regarding it to be an anticipation of phenomenology. (P. 43). But Wild, like all commentators, neglects to perform a study of James's dualisms, resulting in a serious omission in the critical literature, an omission which the present paper may serve to correct.

CHARLES PEIRCE AND THE FIRSTNESS OF PROCESS

SANDRA B. ROSENTHAL
Loyola University, New Orleans

In any discussion of Peirce's philosophy, Firstness is usually the most neglected of his categories. This, however, is not due to any unique clarity of the category of Firstness. Indeed, Isabel Stearns remarks that "Firstness is without any doubt the most elusive of Peirce's categories."[1] And, as Boler accurately summarizes the plight of Firstness, it "is certainly the least clear of the categories and the one that receives the least attention."[2] In the following essay, the focus of attention will be on the category of Firstness as it functions in Peirce's metaphysics.

Although the characterization of metaphysical Firstness which will emerge from the following discussion will differ quite markedly from the general trend of the commonly accepted interpretations, its justification will be threefold. First, it will be seen to follow directly from Peirce's statements concerning epistemological issues in accordance with a method Peirce himself advocates. Secondly, it will provide a more well-integrated category of metaphysics. And, finally, this integrated metaphysical category of Firstness which follows from Peirce's epistemology will be seen to fit in more adequately with and provide a unifying factor in the general spirit of Peirce's pragmatic philosophy. Pragmatists in general break sharply with the older static conceptions of philosophy and see this as a universe in the making; in their philosophies unchanging absolutes give way to process. The present interpretation is intended to better integrate the

[1] Isabel Stearns, "Firstness, Secondness, and Thirdness," *Studies in the Philosophy of Charles Sanders Peirce*, ed. Weiner and Young (Cambridge, Massachusetts: Harvard University Press, 1952), pp. 196–197.

[2] John Boler, *Charles Peirce and Scholastic Realism* (Seattle: University of Washington Press, 1963), pp. 122–123.

category of Firstness with the spirit of pragmatism and the concept of process as it is found in Peirce's metaphysical vision of cosmic evolution.

Before examining Peirce's metaphysical category of Firstness it will be necessary both to distinguish and then interrelate the metaphysical and phenomenological aspects of his categories. As Douglas Greenlee quite adequately characterizes the difference, the categories as metaphysical or ontological are "modes of being", while the categories as phenomenological[1] are "classifications of all that is in any way present to mind in experience."[2] This, then, is the distinction between the two aspects of the categories. However, as Greenlee further notes, since "what is present to the mind is not necessarily mental, there is so far no disparity between the categories as ontological and as phenomenological" (p. 52). Moreover, as Peirce's own statements will be seen to indicate, not only is there no disparity, but rather the categories as phenomenological provide the key to understanding the categories as ontological. That Peirce intends such a relationship between the two types of categories is evidenced in his assertion that "The premisses of nature ... though they are not the perceptual facts that are the premisses to us, nevertheless must resemble them in being premisses. We can only *imagine* what they are by *comparing* them with premisses for us."[3] And, as he explicitly relates this general position to Firstness, "As premisses they involve qualities" (*CP*, 5, 119). Perceptual experience, then, indicates that the category of Firstness, as a "mode of being", implies qualitative richness. And, as Peirce further indicates, "No sensation nor sense faculty is requisite for the possibility which is the being of the quality"

[1] It is the categories as phenomenological that enter directly into Peirce's discussions of epistemological issues.

[2] Douglas Greenlee, "Peirce's Hypostatic and Factorial Categories," *Transactions of the Charles S. Peirce Society*, IV (1968), 51–52. (The distinction made by Greenlee which is discussed above is not the distinction indicated by the title of his article.)

[3] Charles Sanders Peirce, *Collected Papers*, Vol. I–VI, ed. Hartshorne and Weiss (Cambridge, Massachusetts: The Belknap Press of Harvard University, 1931–1935); Vol. VII, VIII, ed. Burks (Cambridge, Massachusetts: Harvard University Press, 1958). 5.119 (Hereafter cited in parentheses in the text preceded by abbreviation "*CP*" and using the conventional two-part number).

(*CP*, 1. 422). Metaphysical Firstness, then, would seem to involve both real possibility and real qualitative richness. And, taking Peirce at his word, it is with Firstness as phenomenological that we should begin in any attempt to better understand this qualitative richness and real possibility.[1]

Murphey gives a concise characterization of Firstness in its phenomenological aspect, stating that:

> A First is not the same as what is usually called a percept ... which has a structure and which combines a number of sense qualities. A pure First ... is simple and devoid of structure. But every percept has a First which is the single impression created by the total ensemble of its elements. Moreover, if a single sense quality of a percept is prescinded from all the rest and is considered by itself, such a quality is a First.[2]

Peirce most succinctly expresses such a characterization in his brief but highly illuminating discussion of quale. As he states, "There is a distinctive quale to every combination of sensations so far as it is really synthesized ... (*CP*, 6. 222). But, as he goes on to note, "Each quale is in itself what it is for itself, without reference to any other ... Nevertheless, comparing consciousness does pronounce them to be alike. They are alike to the comparing consciousness, though neither alike nor unlike in themselves" (*CP*, 6. 224). Peirce clarifies the meaning of this latter assertion in another passage. Taking the example, "Yesterday I saw a blue color; and here is a blue color," he notes that "some beginner may object that they have both blueness in them; but I reply that blueness is nothing but the idea of these sensations and of others I have had, thrown together and indistinctly thought at once" (*CP*, 7.392). Thus, we cannot compare presentations in terms of the quality, "blueness", for the repeatable quality

[1] In "A Pragmatic Concept of the Given," *Transactions of the Charles S. Peirce Society*, III (1967), 74–95, I have given a somewhat detailed analysis of the role of Firstness in perceptual experience. In the present essay only a brief sketch of this role is presented in order to reveal and then explore its implications for an interpretation of the metaphysical role of Firstness. Further support for the brief sketch presented here can be found in that paper.

[2] Murray Murphey, *The Development of Peirce's Philosophy* (Cambridge, Massachusetts: Harvard University Press, 1961), p. 307.

is itself dependent upon the assimilation of past and present presentations.[1]

It would seem, then, that Firstness, as qualitative immediacy, must be considered in two ways. First, it is a repeatable and recognizable quality or quale – that which has been seen before and may be seen again.[2] But, secondly, this repeatability is itself a product of the synthesizing activity of consciousness acting upon unique qualia. This more fundamental level of unique qualia gives significance to Peirce's statement that "Firstness is predominant, not necessarily on account of the abstractness of that idea, but on account of its self-containedness. It is not in being separated from qualities that Firstness is most predominant, but in being something peculiar and idiosyncratic (*CP*, 1.302). Indeed, it is precisely such a characterization which leads Peirce to speak of Firstness in terms of qualities of feeling (*CP*, 5.444), for what are the characteristics of feeling according to Peirce? "There is no resemblance at all in feeling, since feeling is whatever it is, positively and regardless of anything else, while the resemblance of anything lies in the comparison of that thing with something else" (*CP*, 1.310).[3]

This level of unique qualitative immediacy is the important level for our present purposes, for if qualia are unique in the most primitive experience of them and if it is "comparing consciousness" that makes them repeatable, then Firstness in its metaphysical aspect would seem not to indicate any sort

[1] This assimilation by which comparing consciousness "produces" repeatable qualities is similar to what Professor Lee calls 'proto-generalization'. Harold N. Lee, "Suggestions Toward a Contemporary Epistemology," *The Southern Journal of Philosophy*, II (1964), 93.

[2] The awareness of qualities as repeatable and recognizable qualia is still a more primitive epistemological level than that of the awareness of qualities as objective properties indicating possibilities of future experiences. The only type of possibilities of future experiences inherent in repeatable qualia is the possibility of repetition. Here, however, it should be noted that qualities as qualia and qualities as objective properties are not meant to be numerically distinct, but epistemologically distinct. They represent different levels of interpretation.

[3] To think of feeling as used by Peirce in terms of psychology is to be misled by a word, for as Peirce himself emphatically states, "If by 'psychology' we mean the positive or observation science of the mind or consciousness ... psychology can teach us nothing of the nature of feeling, nor can we gain knowledge of any feeling by introspection, for the very reason that it is our immediate consciousness" (*CP*, 1.308).

of determinate repeatables. To allow the repeatability of qualia to lead to a metaphysics which gives an independent ontological status in any sense to determinate repeatables is completely to ignore this most basic mode of Firstness as it enters into experience.

But even the unity of a unique "self-contained" quale has lost some of the original diversity, for as Peirce notes: "That very same element of experience, the quale-element, which appears upon the inside as unity, when viewed from the outside is seen as variety" (*CP*, 6.236). And, again, "No unity can originate in concentration ... but any unity there (sic) was there already may in that way, be many times intensified" (*CP*, 6.227). And, since an uninterpreted quale is itself a "synthesis of sensations" then surely at this level also, what is seen on the inside as unity will be seen on the outside as variety. Here, however, it is crucial to note that although Peirce uses the terms 'impression' or 'sensation' quite often, he explicitly indicates both that there are no first impressions of sense (*CP*, 5.213; 7.465) and that when he does use the term 'impression' it is used as a limiting concept to indicate the boundary of consciousness.[1] Thus, the concept of a synthesis of impressions or a "total impression" merges with the concept of the point of organism environment interaction or the concept of strands of stimuli as seen. Just as recognition unifies diverse qualia, so qualia unify diverse stimuli.[2] And, just as the unifying function of recognition must have some basis upon which to work – however vague this basis may be – so the unifying function of the "production" of qualia must have some objective basis, no matter how vague, upon which to work.

Thus, the stimuli presented for organization, though not fully structured, have a vague character of some sort. In short, the repeatable and recognizable qualia have a somewhat indeterminate but objective basis in the diverse qualitative stimuli. In more Peircian terms, the First Category is

[1] Murphey, *op. cit.*, p. 415; Appendix, Draft 4 of Peirce Manuscript.

[2] The terms 'qualia' and 'stimuli' are not meant to indicate a numerical distinction but rather a logical or epistemological distinction: qualitative richness as grasped by consciousness and as independent of consciousness respectively. Peirce holds quite emphatically to a theory of direct perception. (See *CP*, 5.56.)

applicable to being (*CP*, 1.487). It is to Firstness as applicable to being that we will now turn.

As a preliminary step in discussing metaphysical Firstness, certain terminological confusions must be clarified. Peirce, in his writings, uses the term 'possibility' to characterize not only Firstness but also Thirdness. And, as an added confusion, he indiscriminately interchanges the terms 'possibility' and 'potentiality'. Peirce's dual use of the term 'possibility' is quite understandable, for since Secondness comprises the domain of the actual, the possible, in a broad sense, must include both Firstness and Thirdness. What Peirce means, however, can be clarified in the light of a few illuminating statements. He holds that "A quality is how something may or might have been. A law is how an endless future must continue to be" (*CP*, 1.536). Again, at times he characterizes his three categories of being as possibility, actuality, and destiny (*CP*, 4.547). Finally, he states that "Generality is either of that negative sort which belongs to the merely potential, as such, and this is peculiar to the category of Firstness, or it is of the positive kind which belongs to conditional necessity, and this is peculiar to the category of law" (*CP*, 1.427). What the contextual meaning of the various pairs of terms used to characterize Firstness and Thirdness indicates in each of these examples is that Firstness involves a weaker type of possibility than does Thirdness. The most appropriate terms to distinguish the possibility involved in each of the two categories would be 'possibility' to indicate the First category, 'potentiality' to indicate the Third category. However, because Peirce indiscriminately switches back and forth between these two terms, the present essay would often be using one term precisely where Peirce is stressing the other term. To avoid such confusion, the terms 'negative possibility' and 'positive possibility' will be used to characterize the possibility involved in Firstness and Thirdness respectively, though what will be meant by these terms is roughly the distinction between mere possibility and potentiality, or, in Peirce's terms, the difference between "a mere may be" (*CP*, 1.304) and a "would be" (*CP*, 2.664). Furthermore, this terminology has an advantage in its own right, for it will indicate clearly the relationship between possibility and generality.

The term 'generality' must be clarified because it also serves a dual function. As Buchler notes, Peirce means by the general the opposite of the singular.[1] Since the singular belongs to the category of Secondness, generality must, in a wide sense, characterize both Firstness and Thirdness. As indicated above, Peirce calls the generality of Firstness negative generality and the generality of Thirdness positive generality. The meaning of these characterizations, however, can best be approached indirectly.

When the category of Firstness was discussed from the perspective of Peirce's theory of perception, it was indicated that Firstness as ontological would be characterized by diverse qualitative stimuli. However, to understand the difficulties that arise in examining ontological Firstness, one further statement made by Peirce in connection with perception must be cited again here. Though Peirce states that the quale element which appears on the inside as unity appears on the outside as variety (*CP*, 6.236), he adds that "no unity can originate in concentration ... but any unity there (sic) was there already may in that way, be many times intensified" (*CP*, 6.227). Thus, the ontological basis for the experience of Firstness is not merely "pure" Firstness or diverse qualitative stimuli in their aspect of diversity, but rather pure Firstness "overlaid" with some unifying element. And, if some element of unity within the diversity is required for the experience of Firstness in the sense of a unified quale, then this unity itself must be "part of" ontological Firstness. Thus, a further distinction between the element of diversity and the element of unity is necessary if Peirce's characterizations of Firstness are to be understood. This, however, leads straight to the problem of positive and negative generality.

Boler, though recognizing the significance of Peirce's switch from substance to process in most areas, states that "there is still a sense in which Peirce argues as Scotus does

[1] Justus Buchler, *Charles Peirce's Empiricism* (New York: Octagon Books, Inc., 1966), p. 23.

for a real common object."[1] Why is this? "According to Peirce, the commonness of qualities, which interested the schoolmen, is but one form – a degenerate form at that – of real generality." Thus, according to Boler, at one point at least real generality for Peirce indicates "real commonness" or repetition of form in some sense. Boler's argument seems to hinge on the unstated assumption that Peirce's ontological category of Firstness implies repeatable, fully structured qualities. On this assumption, since Peirce declares that Firstness involves generality, the "real generality" of Peirce's position, at this point at least, would be similar to the scholastic concept of the common nature as a "real common object". On this view, then, the degenerate or negative generality of Firstness provides a unifying factor by providing a rigid structuring of determinate repeatables. But questions remain. In what sense is this a negative or degenerate generality? Furthermore, did not Peirce's discussion of phenomenological Firstness lead toward the expectation that the unity of the diverse stimuli would be not a rigid structure of repetition but rather a somewhat indeterminate basis for a rigid *epistemological* structuring of repeatable qualities? However, this latter view appears to run into problems of its own, for it is not readily evident that Firstness as diverse qualitative stimuli can in any way be characterized as general. And, if the present interpretation is to find justification in Peirce's writings, then the real negative generality of Firstness must be accounted for.

What characterizes the general, as opposed to the singular, is the fact that the law of excluded middle does not apply to the general (*CP*, 5.448). It would seem, then, that one could hold the diverse stimuli of the evolving universe, in their diversity, to be general in the negative sense that *no* determination can be made of them. Thus, though it is true that "a triangle in general is not isosceles nor equilateral; nor is a triangle in general scalene" (*CP*, 5.505), yet a triangle in general is triangular, and the generality of triangularity does limit the possible alternatives of further determination.

[1] John Boler, *op. cit.*, p. 158.

However, it would seem that the diverse stimuli, in their diversity, display a negative generality in that they are limited by nothing whatsoever. Peirce's reference to Firstness in this pure sense which emphasizes the qualitative uniqueness of each of the stimuli can be seen from his statement that "I cannot call it (Firstness) unity, for even unity supposes plurality" (*CP*, 2.83).

At this point, however, another problem arises, for the negative generality of Firstness has not accounted for the unifying element required by Peirce. The clue to the nature of this unity is found in Peirce's statement that "The general is seen to be precisely the continuous" (*CP*, 8, p. 279). Generality, then, must involve continuity; hence, the generality of Firstness can only be fully understood when this category is viewed from the aspect of the unity or continuity which pervades it. Here it may be objected that continuity belongs to the category of Thirdness. However, if the general is the continuous, then the negative generality of Firstness must imply a negative continuity which belongs to the category of Firstness rather than Thirdness. The negative continuity of Firstness, like the negative generality, indicates a negative possibility or mere "may-be" which contains no positive possibility or "would-be" and which thus provides no positive range for further determinations. As Peirce states the position, "Firstness is essentially indifferent to continuity" (*CP*, 6.205). Indeed, just as feeling was seen above to refer to that quale element which in its purity can be related to nothing beyond itself, so the negative generality and continuity of Firstness, which forms the cosmological basis for our experience of qualia, can be related neither to what has been nor to what will be; it has no relatedness, it contains no "would-be"; in short, it is a qualitative continuum of negative possibilities, a "substratum" of pure chance. Perhaps this clarifies the meaning of Peirce's statement, usually interpreted as indicating an idealistic metaphysics, that "wherever chance-spontaneity is found, there in the same proportion feeling exists. In fact, chance is but the outward aspect of that which within itself is feeling (*CP*, 6.265.)

Observing the fact that both "abstract qualities" and

"chance variations" belong to the category of Firstness, Greenlee notes that they seem to have in common only the fact that they are neither Seconds nor Thirds and therefore are relegated to the category of Firsts. And, as he notes, it may well be asked why chance variations should not be assigned to some new fourth category.[1] On this view, Firstness seems to have become the systematic dump-heap for that which will not fit into the categories of Secondness and Thirdness. Yet, if Firstness is indeed first, one would expect it to provide the significant starting point for the metaphysical functions assigned to the other categories. If, as is here held, qualitative immediacy and chance variation are intimately related, then the need for some "new fourth category" dissolves and Firstness does indeed become the significant starting point of Peirce's metaphysics, for it indicates the infinitely varied, concrete qualitative richness of a universe in process, the substratum of pure chance within which random activities occur and begin to take on habits.[2]

Peirce's discussion of Firstness as ontological possibility (as opposed to ontological generality) has frequently led to its identification with some type of Platonic essence. This approach may at first glance seem a mere repetition, couched in different language, of the points made above in connection with Boler's analysis. However, this approach brings to light an entirely different aspect of the problem of interpreting Peirce's category of Firstness.

Haas defends this "Platonic" line of interpretation, holding that "The possible seems to include for Peirce the universe of logical possibility or an ideal world. Some of these ideal, logical possibilities occur in the real world also. 'The sensible world is but a fragment of the ideal world.'"[3] And, notes

[1] Greenlee, "Peirce's Hypostatic and Factorial Categories," *op. cit.*, 55, 58. It should be noted that the present interpretation does not characterize the qualitative aspect of metaphysical Firstness as abstract, as does Greenlee, but rather as an infinitely rich concreteness. In "A Pragmatics Concept of the Given," *op. cit.*, 86–87, I have attempted to show that Peirce's characterization of Firstness as abstract refers to an epistemological function, not an ontological status.

[2] The view that the universe evolves from chance toward uniformity through the habit taking tendencies of originally random flashes of activity is explicitly developed in Peirce's cosmology. See *CP*, 1.412–1.414.

[3] W. P. Haas, O.P., *The Conception of Law and the Unity of Peirce's Philosophy* (Notre Dame, Indiana: University of Notre Dame Press, 1964), p. 99 (*CP*, 3.527).

Haas, Peirce insists that "the possible is a positive universe of being" (*CP* 8.303).

Peirce, however, offers a clarification elsewhere which places these statements in a quite different light. He notes that "My old definition of the possible as that which we do not know not to be true (in some state of information real or feigned) is an anacoluthon. The possible is a positive universe ... but that is all. Of course, there *is* a general logical possible ... but there is also a possible which is something else" (*CP* 8.308). This possible which is something else is a "positive universe of being". And, this possible as a positive universe of being is the negative possibility of Firstness indicated above. The possibility of the ideal world, of which the sensible world is but a fragment, is not another Platonic world which in some way allows the actual sensible world to participate in reality, but rather is an ideal world of logical possibilities whose structure is dependent upon the intelligence of man. As Peirce states, "It has come about through the agencies of development that man is endowed with intelligence of such a nature that he can by ideal experiments ascertain that in a certain universe of logical possibility certain combinations ocuur while others do not" (*CP*, 3.527). Again, Peirce observes that "It is a part of the process of sensible experience to locate its facts in the world of ideals. This is what I mean by saying that the sensible world is but a fragment of the ideal world" (*CP*, 3.527). In short, the ideal world as indicating a realm of logical possibilities within which the actual world must be located is not some realm of metaphysical forms; indeed, it is not a topic for metaphysics at all, but rather belongs to the area of epistemology. The ideal world is the conceptual world of the logically possible or the consistently thinkable within which the facts of experience must be located. To turn such a "conceptual world" into an ontological world is an unwarranted reification which leads to a static conception of the metaphysical possibilities of Firstness.[1]

[1] In "The World of C. I. Lewis," *Philosophy and Phenomenological Research*, XXIX (1969), 589–597, I have attempted to show that a similar confusion between the conceptual and the ontological by critics of Lewis leads to their assertions of inconsistencies in his pragmatic conception of a reality in process.

We have seen that the real possibility of Firstness is a negative possibility which must be carefully distinguished from the positive of Thirdness and from the logical possibility which belongs in the discussion of epistemological issues. The real qualitative richness of Firstness is the richness of diverse qualitative stimuli which "contain" two distinct aspects, an aspect of total diversity and an aspect of somewhat indefinite unity, characterized by negative generality and negative continuity respectively. These two aspects of the qualitative richness are analytically distinct only, and together they constitute a continuum of qualitative diversity which is the very being of the negative possibility of Firstness and the substratum of pure chance.

Thus, there emerges in Peirce's philosophy a metaphysical category of Firstness which is neither a remnant of traditional conceptions of determinate repeatable qualities, nor a remnant of traditional conceptions of eternal Platonic possibilities. Rather, what emerges is a Firstness of process which attributes to process an intimately interrelated set of precisely those characteristics most antithetical to such traditional conceptions.

THE MEANING OF SAMENESS OR FAMILY RESEMBLANCE IN THE PRAGMATIC TRADITION

DAVID L. MILLER
University of Texas – Austin

Harold N. Lee has made a significant contribution to philosophy and for the most part, as I understand them, his conclusions and his defense of them are complementary to American pragmatism. Charles Morris has shown that the writings of the main exponents of pragmatism can be understood best in terms of Peirce's Universal Categories, Firstness, Secondness and Thirdness.[1] Here I shall apply these categories to a very limited problem, namely that of explaining the meaning of identity and sameness in the pragmatic tradition, and more specifically to a problem that arises in connection with G. H. Mead's contention that the significant symbol and universality emerge when an individual, by his own gesture, is able to evoke in himself the *same* response that he evokes in another. (Since the response made by P_1 belongs to P_1 and not to P_2 what can "same" mean?). This task will require an account of the experience of particulars or individuals as well as the meaning of universals.

I shall defend the claim that although particulars are experienced and the category of Firstness applies to them, we cannot say anything about them nor think about any one of them in isolation from other particulars. Each particular has a quality about it that is unique and ineffable and as a consequence universals cannot reside in particulars, nor can any two particulars be existentially identical. Further, as I hope to explain, no two particulars can have existentially identical properties or qualities. This means that if two or more particulars are alike or belong to the same class, it is not due to Firstness or to a property or aspect which presumably

[1] Charles Morris, *The Pragmatic Movement in American Philosophy*, (New York: George Braziller, 1970), pp. 118 ff.

could be lifted out of one particular and exchanged for that "same" property in another particular without making a difference in the particulars. Commonness or universality is not to be found in particulars as particular and Firstness cannot include it, yet Firstness does apply to each particular in isolation from all others.

Secondness applies to the interaction of particulars with each other, and there is no particular, as far as can be known, that does not interact with some other particular or particulars. Still Firstness plus Secondness is not enough to give us that similarity or sameness of which we speak when we say two or more particulars are alike or similar, or the same in some respect or respects. We must look for similarity or family resemblance or what I shall call Functional Identity in Thirdness. Consequently, particulars are similar to each other or belong to the same class only because they interact in a lawful way (Thirdness). Without Thirdness particulars would remain ineffable and unintelligible, and similarity, universality, and functional identity would not apply to anything. In brief, the locus of universality is found in the lawful way in which particulars function as they interact with each other. If two or more particulars interact with other particulars in accordance with a specified law, then they are said to be similar or functionally identical or alike. Without law, the general, habit, or universality (which cannot be found in Firstness or Secondness or in bare quality and possibility), no two entities can be said to be similar in any respect, and each would remain qualitatively different from every other and ineffable, since effability requires cognition and universality.

If a particular can be said significantly to be identical with itself, that is because it is simply itself and not another thing, and in fact there is no sense in which any property, aspect or quality of any given particular is existentially identical with that of any other particular. This means it is nonsense to say that two particulars are similar or alike because each has a property that is existentially identical, or each has existentially the "same" property, aspect, or quality. Similarity or universality is not to be found in existential identity, but only in functional identity, and two particulars are functionally identical or are the same insofar as both interact with other particulars according to a specific law.

I do not want to suggest that particulars do or can in fact exist (or have substantial or existential identity) apart from interacting with other particulars in a lawful way. Each category by itself is an abstraction since in fact whenever experienced particulars become or are intelligible to us all of the categories are required. Still quite often we experience a particular that is not at the time intelligible, and the function of further experience plus reason is to make it intelligible or rational, and it becomes so only when we find the law according to which it behaves, and this involves classifying it as well.

In accordance with the basic tenets of pragmatism: Beginning with an experience of ineffable particulars, the function of reason is to establish habit, to arrive at generality, law, universality, or to arrive at a knowledge of how particulars react with each other. Experience itself is interaction (or transaction) between an organism and its environment.

George Herbert Mead's greatest contribution to pragmatism and to twentieth century philosophy consists in his discovery that (1) minds and an awareness of habit or Thirdness emerges out of a social process in which the action (behavior or responses) of separate individuals are phases of social acts of adjustment usually beneficial to each participant in the act and (2) the self emerges when the individual is able to respond to its own behavior, not in isolation from the behavior of other participants in a social act, but as a phase of the more inclusive act. But being aware of one's own behavior as a phase of a social act requires that one be aware of the habitual, orderly, or lawful response which one's gesture (or behavior) evokes in an other and hence (3) an awareness of Thirdness, universals, emerge in the social process when participants in social action can evoke in themselves (implicitly if not explicitly) the *same* responses (or functionally identical responses) that they evoke in an other.

Our task now is to explain what it means to say that a given individual can, by his own gesture, evoke in himself the same response that he evokes in an other.

Functional Identity or Similarity and Existential Identity

Each particular is what it is and therefore neither it nor any of its aspects can be existentially identical with any other particular or its aspects. And "In pure Secondness, the reacting correlates are *Singulars*, and as such are *Individuals*, not capable of further division."[1] How can we make sense, then, out of Mead's claim that a person can take the role of another person by evoking in himself the "same" response (by his gesture – a significant symbol or a language gesture) that he evokes in another participant in the social act? Clearly your response is yours and mine is made by me, and they are existentially absolutely different, and they usually exist in different places and at different times.

The answer to the above stated question is that, if P_1 is aware of the response his gesture will elicit in P_2, he is first of all aware of a habit or a habitual response, and this habitual response is the general form of a response or it is the lawfulness or orderliness with which that gesture evokes the customary response in the social process. This means in effect that P_1, in making a significant gesture, is acting according to a rule or a custom that emerged out of the social process. P_1, then, must be aware of Thirdness as applied to thc social adjustive act.

Mead offers a behavioristic account of the significant symbol,[2] or he explains conception and thinking (or what traditionally has been termed "mental" and "subjective", in contrast to overt behavior and the objective) in terms of overt behavior, but he does not reduce the symbolic process to physical or mechanical behavior. When one is aware of the response his gesture will evoke in an other, he is by that fact aware of a universal. He is aware of a custom or he is aware of a habit. A habit is Thirdness, it is what applies to an indefinite number of particular instances even as the request "close the

[1] Charles Sanders Peirce, *Collected Papers*, edited by Charles Hartshorne and Paul Weiss. (Cambridge: Harvard University Press, 1934), V, 70.

[2] See G. H. Mead, "A Behavioristic Account of the Significant Symbol," *Journal of Philosophy*, XIX (1922), 157–63.

door" is appropriate and applicable to an indefinite number of cases.

Thus to evoke in one's self the *same* response that is evoked in another cannot mean that these responses as particular events are existentially the same or identical. Rather, the response that is evoked in both P_1 and P_2 is the Universal (the symbolic). It is an awareness of the form which applies to each existentially different response that falls under a given habit or rule or General. It is the General, Thirdness that is shared by both P_1 and P_2 when they share the meaning of a given gesture. And when P_1 and P_2 share the meaning of a gesture, that meaning is universal. In that case there is only one meaning, and it is experienced by both P_1 and P_2.

Universality, Thirdness, is in fact sharable, and that is what makes communication by use of language symbols possible. When the speaker knows the meaning (or has or is aware of the meaning) of his gesture, then he must be aware of that meaning as had also by the listener, assuming the latter understands the meaning as intended by the speaker. Now $P_{1's}$ awareness is existentially and absolutely different from $P_{2's}$ awareness, but that of which they are aware, the habit, is one and the same, or it is a particular habit, but it must be general to be so, i.e., applicable in principle to many instances. The particular response evoked in P_2 by $P_{1's}$ gesture is experienced by both P_1 and P_2, and there Firstness is involved. Still it is known only as an instance of an act (a response) that follows in accordance with a rule or a custom shared by both participants.

Habit, the Universal and Thirdness

Since Thirdness is a category, we are not obligated to explain why, say, when a chess player makes moves in accordance with rules, these rules are the same in each instance. It ought to be clear that each move is a particular event that transpires and that rules endure at least long enough so that particular, passing moves can be instances that conform to them. It is impossible in principle that there be a rule, a habit, a law, or a universal that is applicable to only one particular instance or to one particular only once. Thirdness, in so far as it is applicable, entails particulars, and particulars, in so far as they are intelligible, entail Thirdness. Or we can say that functional identity and existential identity are intelligible only as related to each other.

Mead, in agreement with Peirce, placed special emphasis on habit and the formation of habits. Both agreed that thinking, if successful, results in establishing a new habit, and Mead defends the view that customs are socialized and institutionalized habits, and that if the individual is aware of such a habit he is aware of the form of acts that can be performed by different participants of the group who are also participants in a social process. Mead shows also that selves and minds emerge when individual participants in the social process are able each to know the meaning of a given gesture (significant symbol or language gesture), and that the meaning of a gesture, in so far as one is aware of its meaning, is universal. The universal is shared or sharable by other participants. It is, we may say, the custom, law, or habit which applies to particular responses evoked when a given gesture is used to evoke a response.

According to Mead individual minds, then, are built up out of Thirdness or universals. At first the child acquires habits and customs unwittingly, and in its early stage of development it is not aware of what it is doing, it is not aware of habits some of which it shares with others. If the child, or any of us for that matter, could carry on the process of adjustment effectively by impulse or previously established habits, that is, if the process of adjustment were not frustrated or impeded at any time, then there would be no need for thinking or for establishing new habits, new universals. According to Dewey,[1] Mead was the first to point out that reflective thinking takes place when an act is cut short of completion or when objects in one's environment have conflicting meanings (arouse conflicting responses) to the individual. At such times action is inhibited and by reflective thinking the form of a new response is created, or a new

[1] See *The Philosophy of John Dewey*, edited by Arthur Schilpp (New York: Tudor Publishing Company, 1939), p. 26.

meaning, a new universal emerges. Although Mead did not make clear the difference and the relationship between existential identity (of particulars – Firstness) and functional identity (similarity of particulars – Thirdness) this distinction and relationship is implicit in all of his writings. Still it is imperative that we explain what it means to say that by his own gesture P_1 can evoke in himself the *same* response that he evokes in P_2 who is another participant in the social act of adjustment. But we must explain also that by use of language gesture one can create new meanings, and that therefore new universals or sharable responses may be created by individuals through reflective thinking. The answer in abbreviated form is simply that the *same* response refers to Thirdness, and for two or more overt responses (which are existentially different) to be functionally identical or similar or the same, they must be subsumable under a Universal, a habit that is in principle applicable to an indefinite number of responses or acts.

Consequently when P_1 makes a request of P_2, and if both understand the meaning of that request, then each experiences, not simply the particular response that is evoked here and now, but the form of the response, or that which is communicable from one individual to another by language gestures, or a sharable response which is *ipso facto* universal. The particular situation of which the social process of adjustment is a part will determine precisely how the Universal is applied. A habit, a universal, has no application apart from these particular circumstances each of which has a tang of novelty (Firstness) about it. "Close the door" cannot have the same meaning to both the requestor and the requestee in any particular instance unless that meaning is general or universal. In other words, it is because "close the door" has a meaning that is universal (shared) that it can be applied and is effective in a particular situation, i.e., have the effect intended by the requestor. By virtue of the relationship between Firstness and Thirdness, it follows that no statement can have universal meaning unless it is applicable, at least in principle to particular instances.

Since, according to Mead, individual minds are built up out of Thirdness, there can be no private language nor can the contents of mind be what traditionally has been called "subjective content". Rather, individual minds or the perspective of the individual, emerges out of a group perspective or a common perspective, and the function of creative intelligence (which is carried on by individuals only) is to furnish new meanings that are sharable by the community and are, therefore, directed toward Thirdness.

Inorganic Behavior and Organic Behavior

I have indicated above that acts may be inhibited or, as Mead would say, the individual may be prevented from fulfilling the consummatory phase of the act. He distinguishes between the perceptual, the manipulatory, and the consummatory phases of an act.[1] Such a distinction applies only to organic behavior, to the life process, or to a relationship between an organism and its environment. If, on the other hand, we consider the "behavior" of a chunk of matter, we find that it is intelligible in terms of its inertia, which is constant, ever the same, with no tendency to complete an act and no tendency to maintain or not to maintain its present state whether in motion or at rest. In this sense inorganic bodies have no environment, and there is no basis for distinguishing between various phases of their action. Hence we can make sense of Mead's contention that organic behavior is teleological (in contrast with inorganic action) and that both the perceptual and the manipulatory phases of the act have meaning only in relation to the consummatory phase. To say an act is teleological is to say it is heterogenous with respect to time, in contrast with inorganic action which is homogenous. Although organic behavior is heterogenous, we identify the first two phases of an act by referring to the consummatory phase, or they have meaning only in relation to it. Whether a person is just beginning to close the door, is half way through or is nearly finished with the job, we say in each case that he is closing the door. It matters not whether the door is closed by P_1, P_2, or P_3. Each case is an instance of

[1] See especially G. H. Mead, *The Philosophy of the Act*, edited by Charles W. Morris (Chicago: The University of Chicago Press, 1938), Chapter I.

closing the door. And since the one who makes a request stimulates himself to answer the request, e.g. arouses in himself the attitude or tendency to close the door, and since he is aware of that attitude, he has the idea of closing the door. This idea is the beginning of an act of closing the door, and since any phase of the act of closing the door is known in relation to its consummatory phase, his attitude, his idea, though inhibited, is one that answers to a closed door. This attitude is communicable to another by significant symbols, and it will answer to an indefinite number of particulars or to the same particular an indefinite number of times. It is Universal. A language gesture both connotes and denotes. It connotes the universal (the concept), that in principle applies to or answers to particulars which, as a consequence, are said to be similar, or they are functionally identical though existentially absolutely different.

If Mead is correct, then definitions, meanings, concepts are not to be found in paradigm cases or in particular instances that are pointed out to one, even as a habit cannot be sensed in a particular manifestation of it. Rather, an awareness of a habit presupposes (entails) a social process in which the individual is able to apprehend his behavior as a phase of a more inclusive social act. In so far as one is aware of Thirdness in the social process he is aware of sharable (if not shared) habitual responses, and responses are transactions between organisms and their environments aimed at the consummatory phases of acts. Awareness of Thirdness is mental but in no case subjective inasmuch as the content of the mental is sharable and therefore universal. Every manifestation of a universal involves particulars in relation to each other (Firstness and Secondness). Thirdness is the order in which transactions between particulars are carried out.

Mead holds that an awareness of an entire social act, the gestural or stimulus phase as well as the response phase, enables one to get control over the response by getting control over the stimulus or gesture. And in many instances this means the creation by reflective thinking of new kinds of responses, new meanings, new universals. Thus minds enter into the creative advance of nature and objects take on new meanings in so far as they answer to these universals, for the meaning an object has cannot be stated apart from universals. And Mead, as well as James, contends that objects have meaning to us only in relation to manipulation or how we behave with reference to them. This implies that our knowledge of the laws of inorganic bodies must be such that they can be used by us or that by our behavior we can apply laws of nature and thereby predict or control the order of events. The fact that men can perform experiments and thereby test hypotheses is evidence enough that laws of nature fit in with the intelligent behavior of men.

Although Mead said very little about Pierce's Universal Categories, I agree with Morris that his thinking is altogether in line with them. He emphasized what he called the exceptional event which serves as a basis for the construction of new laws and habits. The exceptional event is experienced as Firstness, and through further experience and experimentation we find Thirdness or simply the lawful way in which it reacts with other particulars.

Mead insisted, in contrast with, say, British empiricists, that the particular as particular cannot be known, though it can be experienced and such experience is confined to Firstness. He emphasized that particulars can be known only in relation to what he called a system. E.g., the mass or inertia of a particular body can be measured and known only by bringing in other bodies and noting the lawful way in which they interact. The mass of one body may be equal to the mass of another, but here equality does not mean existential identity, but rather functional identity. They are functionally identical because they conform to the same law, or one can be substituted for another by applying a given law or practice, such as balancing one or the other chunk of matter with another given chunk. Still our knowledge of the laws of inorganic matter must be stated in terms of instruments and systems of weights and measures that have been devised by men, in terms of concepts that grew out of social behavior. This being the case it is clear that our knowledge of Thirdness as applied to what are usually called the laws of nature is intimately and necessarily related to human habitual responses that have been institutionalized and of which we are aware. Such responses are shared and they (that is, the Thirdness that attaches to them) can be communicated from one individual to another. For this reason we are justified in saying that minds and selves are made up out of contents that are objective, out of Thirdness that is not a mere collection or aggregate of particulars, but rather the universal, the systematic order in which particulars interact.

To be aware of a habit (Thirdness) is to be aware of something that transcends any of the particular instances subsumed by it. It involves something other than Firstness and Secondness, which are experienced in particulars but as such are not known. If Mead is correct in his well supported claim that minds are built up principally out of an awareness of Thirdness, and that such awareness can emerge only in a social process of adjustment in which the participants can take the roles of other participants or enter into each others perspectives, then he has shown, in accordance with the basic tenets of pragmatism, that Thirdness, wherever it applies, can be understood and communicated from one person to another only in so far as the communicants are participants in social behavior, and that an understanding of Thirdness as applied to "laws of nature" depends upon the principle of sociality. Through the behavioristic psychological approach he has solved the problem concerning the nature and locus of universals, the nature of reflective thinking and communication by use of language gestures.

If communication depended upon eliciting in another an existentially identical response, it would be impossible. But, by resorting to a distinction between the three Peircean categories, we are justified, short of begging the point in question, in saying that the individual can, by his gesture, evoke in himself the same response that he evokes in the other, and that sameness refers to Thirdness, that form that applies to an indefinite number of particular or particular instances.

Also Mead has shown how to approach many problems that have arisen in present day philosophy, such as the problem of private language, solipsism, the priority of one's own case, the priority of the other's case, subjectivism, the mind-body problem, and the problem of the relationship between the given, or what is in immediate experience, and knowledge. He has also contributed immensely to the pragmatic tradition by explaining precisely why minds and the contents of minds cannot be explained apart from the social activities of individuals, activities that constitute transactions between men and their environments.

Mead's Principle of Sociality, which he outlines but does not fully develop in *The Philosophy of the Present*, is aimed to show how in the evolutionary process new universals, new Thirdness arise. His contention is that the same general format that applies to the emergence of new universals due to reflective thinking applies to the emergence of new laws, new systems throughout nature. This is not to be interpreted to mean that Mead is finally an idealist, but rather that mind is a natural phenomenon: "... the culmination of that sociality which is found throughout the universe."[1] Mind is the source of new Thirdness, and this is possible because particulars which are existentially absolutely different from each other nevertheless lend themselves to many different kinds of interactions with each other, interactions that are lawful, many of which would not have taken place apart from the intervention of reasoning carried on by individual members of society.

[1] G. H. Mead, *The Philosophy of the Present*, edited by Arthur E. Murphy (Chicago: The Open Court Publishing Company, 1932), p. 86.

PRAGMATICS AND DEFINITE DESCRIPTIONS

JOHN A. BARKER
Southern Illinois University at Edwardsville

One philosophical logician will take the current system of formal logic itself as the firm base from which to direct all his ordering operations against the unruly tribes of language, all his explanatory manoeuvres among the diverse phenomena of speech. Another will start – as far away as possible from this base – by examining those phenomena themselves in all their particularity and detail. Will the true order be achieved when the adherents of both approaches are folded in a single party, when we have a theory of meaning firm enough and rich enough to accommodate in a single, unified account both the simplicity of logical laws and the diversity of linguistic phenomena? Perhaps it will – if there is such a thing as the true order.[1]

1. If we adopt the dual objective of attempting to provide a precise, logical framework for analyzing discourse while maintaining a close liaison with ordinary discourse, we find that existing theories of definite description seem not completely satisfactory.[2] For example, of the best known views, Russell's provides many of the requirements for a formal framework, but does not capture the apparent referential role of definite descriptions in ordinary discourse.[3] Strawson's view, on the other hand, is a referential interpretation which postulates that statements involving definite descriptions with no existing referents are truth-valueless; hence Strawson's statements are not amenable to treatment in standard

[1] P. F. Strawson, *Philosophical Logic* (London: Oxford University Press; 1967), p. 2.

[2] For collections of major articles on definite descriptions, see e.g., Jay Rosenberg and Charles Travis, eds., *Readings in the Philosophy of Language* (Englewood Cliffs, N.J.: Prentice Hall, 1971), and Thomas Olshewsky, ed., *Problems in the Philosophy of Language* (New York: Holt, Rinehart & Winston, 1969). For a comprehensive treatment of the main theories, see Leonard Linsky, *Referring* (New York: Humanities, 1967).

[3] Bertrand Russell, "On Denoting" *Mind*, XIV (1905), 479–93, and *Introduction to Mathematical Philosophy* (London: Allen and Unwin, 1919), Chap. XVI.

logics.[1] Those referential views, such as Meinong's, which avoid truth-value gaps by permitting reference to non-existing entities were long ago repudiated by Russell as clashing with "a robust sense of reality."[2] Theories which postulate shifting references and arbitrary "null" referents, such as Frege's view, do not appear to match ordinary discourse very closely.[3]

But even if a *prima-facie* satisfactory referential view could be devised, there is an additional apparent difficulty which seems to destroy all hope of making the view capture the role of definite descriptions in ordinary discourse. Donnellan has presented strong arguments to the effect that in ordinary discourse definite descriptions are *ambiguous*, sometimes being used referentially, but at other times being used "attributively," to state something about *whatever* fits the description.[4]

[1] P. F. Strawson, "On Referring," *Mind*, LIX (1950), 320–344; "A Reply to Mr. Sellars," *Philosophical Review*, LXIII (1954), 216–231; and "Identifying Reference and Truth-Values," *Theoria*, XXX (1964), 96–118. The latter two articles provide qualifications to Strawson's view which make the above remark an oversimplification. The theory developed in the present essay is in many respects a Strawsonian view modified to prevent truth-value gaps, to permit reference to non-existents, and to take account of considerations advanced by such authors as Charles Caton, "Strawson on Referring," *Mind*, LXVIII (1959), 539–544; Keith Donnellan, "Reference and Definite Descriptions," *Philosophical Review*, LXXV (1966), 281–304; Richard Gale, "Strawson's Restricted Theory of Referring," *Philosophical Quarterly*, XX (1970), 162–65; Frank Leavitt, "On Strawson's Revised Position on Identifying Reference," *Theoria*, XXXV (1969), 64–70; and Leonard Linsky, *op. cit.*

[2] Alexius Meinong, "The Theory of Objects" (trans. by Isaac Levi, D. B. Terrell, and Roderick Chisholm) in *Realism and the Background of Phenomenology*, Roderick Chisholm, ed. (Glencoe, Ill.: The Free Press, 1960) pp. 76–117. For Russell's discussion of Meinong's views, see *Introduction to Mathematical Philosophy*, *op. cit.*, p. 168 ff. For a view similar to Meinong's, see Hugh MacColl's series of articles in *Mind* XIV (1905).

[3] Gottlob Frege, "On Sense and Reference" in P. Geach and M. Black, eds., *The Philosophical Writings of Gottlob Frege* (Oxford: Basil Blackwell, 1960).

[4] Keith Donnellan, "Reference and Definite Descriptions", *op. cit.* "A speaker who uses a definite description attributively in an assertion states something about whoever or whatever is the so-and-so. A speaker who uses a definite description referentially in an assertion, on the other hand, uses the description to enable his audience to pick out whom or what he is talking about and states something about that person or thing" (p. 284).

See also "Putting Humpty Dumpty Together Again," *Philosophical Review*, LXXVII (1968), 203–215. For a treatment of pragmatics which relates closely to Donellan's view, see Robert Stalnaker, "Pragmatics," *Synthese* XXII (1970), 272–289. (The present essay developed from a commentary on Stalnaker's paper presented to the Illinois Philosophy Conference, 1970).

If such an ambiguity exists, a program of providing a logical framework closely matching ordinary discourse appears hopeless.

2. In this paper an attempt will be made to outline a referential view of definite descriptions which introduces no truth-value gaps. The theory will be designed to permit reference to non-existent entities, and will make it possible to assert truths about them (e.g., that they do not exist). Nevertheless, an explication of existence will be provided which will ensure that almost any positive assertion about a non-existent entity will entail its existence, and hence will be false. Moreover, it will be seen that, *normally*, referring to any entity presupposes its existence. Thus, the theory will show that while under certain circumstances non-existent entities can be referred to, identified and denied existence, they cannot (roughly speaking) be truly said to be or to do anything. It will be seen that the practical advantage of being able to make true statements about an entity which turns out not to exist is to be able to collect enough true statements about it to build a case for its non-existence. For these reasons an acceptance of the theory will not entail a loss of one's "robust sense of reality." The theory, while maintaining that definite descriptions are always used referentially, will incorporate the substance of Donnellan's insights without requiring the postulation of an ambiguity affecting definite descriptions in ordinary discourse.

3. Our strategy will be as follows. As our formal framework we shall use standard, second-order quantificational logic, with identity defined in the usual way [two individuals are identical if and only if they have all of their attributes (properties and relations) in common.] We shall not alter the rules for quantifiers, but we shall interpret the individual quantifiers as ranging over "possible" individuals – individuals which may or may not exist. An individual constant, which is to represent a referring expression, can be introduced only if there is some individual (existing or merely possible) which is being referred to by the constant. Similarly, an attribute constant can be introduced only if there is some attribute (property or relation) which it designates. An indi-

vidual will be said to exist if and only if it has at least one attribute.[1] Since there are predicates, such as '. . . is identical with . . .', '. . . is bald-or-not-bald', '. . . is feared', which can be predicated of non-existent individuals, such predicates will be viewed as not designating attributes, and hence will not be represented as attribute constants. As we shall see later, such predicates are to be analyzed in terms of *propositional* operators (e.g., to say x is identical with y is to say x has an attribute *if and only if* y also has the attribute, something which can be true even if x and y have no attributes – i.e. do not exist).

In order to provide a set of concepts applicable to ordinary discourse and capable of facilitating the interpretation of ordinary statements in terms of the formal framework, we shall have to deal with the *pragmatic* dimension of language, that aspect of language involving the relationships of language with its users. In particular, we shall have to concern ourselves with speech acts and speech products. This dimension of language is, as substantial recent investigation has shown,[2] so complex that we shall have to limit ourselves to bare sketches; moreover, we shall focus only on the speech act of *asserting*, discussing *meaning, use, statements*, etc., only in relation to this particular speech act. Thus, we shall not be concerned with such things as questions, commands and promises, nor with the role of definite descriptions in them.

We shall attempt to explicate *asserting* primarily in terms of the concepts of *believing* and of *producing something in*

[1] For somewhat similar treatments of existence, see Henry Leonard, "The Logic of Existence," *Philosophical Studies*, VII (1956), 49–64; Nicholas Rescher, "On the Logic of Existence and Denotation," *Philosophical Review*, LXVIII (1959), 157–180; A. N. Prior, *Past, Present and Future* (Oxford: Clarendon Press, 1967), pp. 161 ff; Nino Cocchiarella, "Some Remarks on Second Order Logic with Existence Attributes," *Nous*, II (1968), 165–175, and "Existence Entailing Attributes, Modes of Copulation and Modes of Being in Second Order Logic," *Nous* III (1969), 33–48. For a critique of Prior's suggestions, see James Tomberlin, "Prior on Time and Tense," *Review of Metaphysics*, XXIV (1970), 57–81. Rolf Schock's *Logics Without Existence Assumptions* (Stockholm: Almqvist and Wiksell, 1968) contains a discussion of the many attempts to deal with existence by means of non-standard quantification theories.

[2] For collections of major articles, see the anthologies cited in note 1 above. For a comprehensive treatment, see John Searle, *Speech Acts: An Essay in the Philosophy of Language* (Cambridge: Cambridge University Press, 1969).

order that such and such happens. What is believed is a *proposition,* which we explicate as a state of affairs (a situation or a happening) considered from the point of view of its being an object of belief. Thus two propositions are identical if and only if the corresponding states of affairs are identical. A (simple) state of affairs, consisting of an individual's (possibly) possessing an attribute, is identical with any state of affairs consisting of an identical individual's (possibly) possessing an identical attribute. A state of affairs may be actual or non-actual, and a proposition "about" the state of affairs is true if and only if the state of affairs is actual, i.e., if and only if the individual actually possesses the attribute. The notions of *proposition, state of affairs* and *assertion* shall be employed to develop conceptions of statements and of definite descriptions without introducing statements which are neither true nor false.

4. The major theses of the paper can be roughly formulated as follows: With a *definite description* defined as any example of 'the so-and-so' expressions in which the speaker implies that there is one and only one individual fitting the description, we shall try to maintain that definite descriptions are always used *referentially* – the speaker in making his statement asserts something *of the unique individual,* and does not assert that there is such an individual. The speaker *always presupposes* that there is a referent for the description, *always presupposes* that the referent fits the description, and *normally presupposes* that the referent exists.

A more specific formulation of the major themes can be accomplished as follows. Consider the following conversation: A, who believes France is presently a monarchy, makes the statement: "The present king of France is bald." B, believing as A, makes the statement: "The present king of France is not bald." C, knowing France is not a monarchy but inadvertently thinking the phrase 'the present king of France' denoted Pompidou (known by C not to be bald), makes the statement: "The present king of France is not bald." Finally D, knowing the truth of the matter, makes the statements: "The present king of France is not bald. He doesn't exist." We shall try to develop conceptions of *statements, assertions, presuppositions* and *reference* which will yield all of the following theses: D is the only speaker whose statements are true; the other speakers' statements are false because the present king of France does not exist. A *presupposes*, but does not *assert*, that the present king of France exists; B also presupposes that he exists. C, unlike A, B, and D, neither refers to nor asserts anything about the present king of France; what C *asserts* is that Pompidou is not bald, something which is true, while his *statement*, by virtue of implying that Pompidou is king of France, is false. A, B and D referred to the same individual, the present king of France, a well-defined individual who can properly be denied existence because France is not presently a monarchy.

5. We begin with an investigation of the speech act of assertion, leading eventually to a definition of statement.[1] To gain a perspective on assertion, we shall briefly discuss an unusual case of lying. Imagine the following situation:

(1) A certain military commander X (unable to communicate directly with two allied commanders Y and Z) moves his troops back to make Y and Z believe that he believes a certain proposition P, say that reinforcements are arriving.

Suppose that X does not really believe P, but succeeds in getting them to believe he does. We can say that X has intentionally misled them about his beliefs; but we would not say that what he did constituted lying to them. Now suppose:

(2) X believes that Y and Z believe that (1) is true.

And suppose X is right in this belief. In this case we would still not say that X has actually lied to them, but the situation is bordering on an actual assertion on his part that P. Suppose finally:

(3) X believes that Y and Z believe that (2) is true.

And suppose X is again right in this belief. Using scarce quotes to indicate a use of 'knowledge' as 'true opinion', we can say that the important dimension added by (3) is X's belief that

[1] This sketch of a theory of assertion borrows heavily from views advanced by H. P. Grice, "Meaning," *Philosophical Review*, LXVI (1957), 377–388, and "Utterer's Meaning and Intensions," *Philosophical Review*, LXXVIII (1969), 147–177; P. F. Strawson, "Intention and Convention in Speech Acts" *Philosophical Review* LXXIII (1964) 439–460; and John Searle, *op. cit.*

they will not only have the "knowledge" that he is intending to do something, but will "know" that he "knows" they have this "knowledge." It seems that we would now have adequate grounds for saying that X has lied to them – he has actually *asserted* that reinforcements are arriving, even though the "language" he used is not of the usual sort.

Let us suppose that the following situation obtains:

(1) a certain individual x (a person) produces an individual y (say, a series of noises or marks) in order that an individual z (another person) believe that x believes that p (a certain proposition).

(2) x believes that z believes that (1) is true.

(3) x believes that z believes that (2) is true.

We propose the following conceptions: x has *asserted* that p; the proposition p is his *assertion*, the product of his speech act of asserting. x is the *speaker*, y the *utterance*, and z the *audience*. x has *referred to* the individuals in the state of affairs corresponding to the proposition p. The *sentence* is the abstract type of which the utterance is the token. The *meaning of the utterance* is the assertion, while the *meaning of the sentence* is the set of possible assertions which could be made by *proper uses* of the sentence, where a *proper use* is a production of an utterance of the sentence in conformity with the speech habits of the speech community. (To *specify* the meaning of the sentence one does not attempt the impossible task of listing the possible assertions, but rather attempts to give general directions for determining which assertions could be made with proper uses of the sentence.)[1]

6. We shall now introduce a conception of *statements* which

[1] The following remarks will perhaps serve to prevent some misunderstandings of the above ideas. If z does believe that (1) and (2) are true, then z may be said to *understand x's utterance*, to *know what x asserted*, and to *know the meaning of x's utterance*. This may be achieved even if z does not know the *meaning of the sentence* x uttered, and even if x's intended effect (that z believe that x believes that p) is not achieved. If x believes there is no chance that this effect could be achieved, then he cannot really intend the effect, and hence cannot really assert that p. This is not to say x cannot assert that p if he believes there is no chance that z will believe p, nor is it to say that x cannot assert that p if he believes that z already believes that x believes that p—x can assert a proposition which he knows; z will not accept, or which he knows z already knows he believes, provided that x believes there is a chance that his utterance will at least "over-determine" that z believe that x believes the proposition.

distinguishes them from assertions by emphasizing their external, audience-oriented character, a character which makes them, unlike assertions, not merely a function of the speaker's intentions. It will be possible for a speaker's statement to contain false information while his assertion is true, and even while all his beliefs are true. To define *statement* we shall need to introduce concepts of *pragmatic implication*, of the *negate* of a sentence, and of *presupposition*.[1] A *speaker* will be said to *imply* a proposition p if and only if a *proper use* of the sentence employed implies that the speaker believes p. The proposition a speaker implies is called a *pragmatic implication*. For example, if a speaker asserts that Bertrand's winning implied Peter's losing by uttering the sentence 'Since Bertrand won, Peter lost', he implies not only the proposition that Bertrand's winning implied Peter's losing, but also the proposition that Bertrand won and the proposition that Peter lost, since proper use of the sentence implies that the speaker believes these propositions. In other words, if the speaker does not believe these propositions, then his production of the utterance of the sentence is not in conformity with the speech habits of our speech community. The *negate* of a sentence used by a certain speaker is another sentence by whose use in the given circumstances a speaker would imply the falsity of the former speaker's assertion, while implying the truth of the other implications. For example, the negate of the sentence 'Since Bertrand won, Peter lost' is the sentence 'Even though Bertrand won, still Peter might not have lost', since by a use of the latter in the circumstances a speaker would imply that Bertrand's winning did not imply Peter's losing, with the pragmatic implications that Bertrand won and Peter lost being maintained. A speaker *presupposes* a proposition p if and only if (1) he implies but does not assert

[1] The present treatment of pragmatic implication and of presupposition profited from the work of such authors as H. P. Grice, "The Causal Theory of Perception", *Supplementary Proceedings of the Aristolelean Society*, XXXV (1961), 121–152; P. F. Strawson, *Introduction to Logical Theory* (London: Methuen, 1952), and *Individuals: An Essay in Descriptive Metaphysics* (London: Methuen, 1959); Roger Montague, "Presupposing," *Philosophical Quarterly*, XIX (1969), 97–110; John Post, "An Analysis of Presupposing," *Southern Journal of Philosophy*, VI (1968), 167–171; G. Nerlich, "Presupposition and Entailment," *American Philosophical Quarterly*, II (1965), 136–152.

p, and (2) there is a negate of the sentence employed such that a speaker using this negate would also imply p. For example, if a speaker asserts that Bertrand's winning implied Peter's losing by uttering the sentence 'Since Bertrand won, Peter lost', then he *presupposes* that Bertrand won and that Peter lost, since these propositions are pragmatically implied both by the given speaker and by a speaker using the sentence's negate, i.e., 'Even though Bertrand won, still Peter might not have lost'.

We are now in a position to define *statement*. The *statement* a speaker makes is the conjunction of the assertion and the presuppositions. For example, the statement made by uttering the sentence 'Since Bertrand won, Peter lost', consists of a conjunction of three propositions: that Bertrand's winning implied Peter's losing (which is the assertion), and that Bertrand won and that Peter lost (which are the presuppositions). If any of these three propositions is false, the statement is false. And any statement which consists of the same assertion and the same presuppositions is the same statement. For example, the statement made by uttering the sentence 'If Peter had not lost, Bertrand would not have won', is the same statement as that made by uttering the sentence 'Since Bertrand won, Peter lost', since the assertion associated with the former (i.e., that Peter's not losing implied Bertrand's not winning) is the same as the assertion associated with the latter (i.e., that Bertrand's winning implied Peter's losing), and the presuppositions (i.e., that Bertrand won and that Peter lost) are common to both statements. It should be noted that the statements made by uttering the sentence (Bertrand won and Peter lost and Bertrand's winning implied Peter's losing' is not the same statement as the previous examples, since even though it consists of the same propositions (and hence has the same truth conditions) it does not consist of the same assertion and the same presuppositions.

7. We shall now attempt to define and characterize definite descriptions, striving for an explication which will permit them to have a consistently referential role without introducing any truth-value gaps for statements involving definite descriptions with no existing referents. Since we want to main-

tain that definite descriptions are always used referentially, and since in some cases expressions of the form 'the so-and so' are not used referentially (e.g., "The whale is a mammal"), we cannot define definite descriptions in terms of their linguistic form alone. But if we define them in terms of their referential function, the question will be begged. We shall employ the notion of a *denoting* use of 'the so-and-so' expressions – a use in which it is implied that there is a unique individual fitting the description. But to avoid ruling out in advance the possibility that a sentence such as 'The present king of France does not exist' can be used to refer to the present king of France, we should be careful not to read 'there is a unique individual' as 'there *exists* a unique individual'. Similarly, in view of our explication of existence as the possession of some attribute, we should not require that the speaker using the definite description imply that the unique individual actually possesses the relevant attribute.

An individual will be said to *exist* if and only if it has at least one attribute. For example, if the present king of France were bald, he would have an attribute, and hence would exist. Since he does not exist, he does not even have the attribute of being king of France. Predicates which can be truly predicated of him are to be "analyzed away" in terms of propositional operators. For example, even though the king does not exist, he is still himself (i.e., identical with himself), since it is true that *if* he had any attributes, *then* he would have these attributes. Similarly, the king is bald-or-not-bald, for it is true that *either* he is bald *or* he is not bald. The king may indeed be feared by some people, since from the fact that they *believe* that he will harm them, it does not follow that he will harm them. Or the king may be thought of, or thought to be such-and-such, since even though he has no attributes someone may *believe, imagine, suppose,* etc., that he exists or is such-and-such. Later we shall discuss interpreting the predication of attributes of *fictional* individuals in such a way that these individuals are not actually said to possess the attributes. Of course, if it turns out that there are predicates truly applicable to non-existent entities which cannot be "analyzed away," then the present explication of existence will prove to be untenable.[1]

To say that an individual *virtually possesses* an attribute is to say that if the individual exists then it possesses the attribute. In this case we are using the phrase 'if ... then ...' in the weak sense of *material implication,* a sense not implying any *connection* between antecedent and consequent. Perhaps a less misleading formulation would be: an individual virtually possesses an attribute if and only if either the individual possesses the attribute or the individual does not exist. In a stronger sense of 'if ... then ...' attributes which were such that a given individual would possess them if it existed (i.e., would not exist if it did not possess them) would qualify as the individual's *essential* attributes.[2] In relation to this notion, the attributes which are virtually possessed by a given individual could be said to constitute a *nominal* essence, an

[1] The suggested analyses of concepts of the "intentional" relations of fearing, thinking of, etc., are not being offered as complete or adequate, but merely as suggestive of the feasibility of this kind of analysis.

It should be noted that the present explication of *singular* existence is analogous to the standard explication of *general* existence – just as the claim that unicorns exist is true if and only if there is at least one individual which possesses this attribute, so also the claim that a given unicorn exists is true if and only if there is at least one attribute which this individual possesses. It is instructive to note that if one is tempted to think that every individual has at least one attribute because, say, it is logically true that there is at least one attribute such that the individual possesses it if and only if the individual possesses it, then using the above analogy we could say it would follow that there is at least one individual which has the attribute of being a unicorn, since it is logically true that there is at least one individual such that the attribute is possessed by it if and only if the attribute is possessed by it. (The present theory's interpretation of individual quantifiers as ranging over "possible" individuals does not undermine the standard explication of general existence; for instance, if there is at least one "possible" individual which possesses the attribute of being a unicorn, it follows that this individual is an existing individual, and hence that there is at least one *existing* individual which is a unicorn.)

[2] For a treatment of this strong sense of 'if ... then ...,' see the author's *A Formal Analysis of Conditionals* (Carbondale, Ill.: Southern Illinois University Monographs, 1969). Possession of this stronger propositional operator would permit us to define a concept of identity strong enough to make it false that a given non-existent individual is identical with any other non-existent individual. With such resources we could also define the *attribute complement* F′ of a given attribute F as the set of attributes which are such that if any individual exists and is not F, then the individual is F′. Thus for all and only existing individuals could it be said that each one either has a given attribute or has its complement. For example, it could be said that speaker B above, in stating that the present king of France was not bald, implied that he had hair.

essence which may turn out to be *merely nominal* rather than *real.*

A phrase of the form 'the F' (in the singular) is a *definite description* in a sentence uttered by a speaker if and only if by uttering the phrase the speaker implies that there is an individual x and a set of attributes G (including but not necessarily limited to the attributes designated by 'F') such that: (1) x virtually possesses G, and (2) for any y, if y possesses G, then y is identical with x. The individual x (if there is such an individual) is called the *denotatum* of the definite description, is said to *fit the description*, and is said to be *characterized* by G, which is called the *concept* of the individual. For example, if a speaker asserts something by uttering the sentence 'The table is not brown,' the phrase 'the table' is a definite description since a proper use of the utterance implies that the speaker believes there is an individual and a set of attributes, say, the attribute of being a table in his office, such that the individual at least virtually possesses the attribute and no other individual possess it. The relevant phrases in the following sentences would normally be definite descriptions: 'The sole survivor is (is not) Nixon', 'The winner of the race did (did not) receive a prize', 'Bertrand does (does not) fear the president of England', 'The president of England does (does not) exist'. In contrast, the relevant phrases in the following sentences would normally not be definite descriptions: 'Nixon is (is not) the sole survivor', 'The whale is (is not) a mammal', 'Happy is the man whose heart is pure'.[1]

8. Are definite descriptions always used referentially? In a typical situation a speaker who made the statement "The table in my room is brown" should, it seems, be taken to have asserted *of the table* that it was brown, rather than to have asserted that there was one and only one table in his room, and that it was brown. Similarly, if the speaker made the statement "The first U.S. president of the 21st century will be a Democrat," he should, it seems, be taken to have asserted *of that president* that he will be a Democrat, rather than to

[1] For a discussion of the role of definite descriptions from a linguistic point of view, see Zeno Vendler, *Linguistics in Philosophy* (Ithaca, N.Y.: Cornell University Press, 1967), Chap. II.

have asserted that there will be someone who will be president of the U.S. first in the 21st century and that he will be a Democrat. Moreover, as the term is ordinarily used, we should, it seems, take the speaker to have *referred to* that table and to that president. Our strategy will be to try to show that this "common-sense" view of the role of definite descriptions can be defended against the difficulties which have led many philosophers to abandon it.

The first main difficulty can be aired as follows. Suppose the speaker only *thought* there was a table there – how can he be referring to anything? In such a case we can say simply that the table does not exist, rather than saying that the speaker did not refer to anything. The speaker's statement is false; nevertheless, since if we asserted its falsity by making the statement "The table is not brown," we would imply that the table existed, our assertion of its falsity should be accomplished in some other way (e.g., "There is no such table" or "The table is not brown; it doesn't exist.") The second closely related difficulty is the following. Suppose there exists more than one such table – how can reference be taking place? At this point we should raise the question whether there is an object (existing or merely possible) *of which* the speaker asserted that it was brown. If not, then the speaker is not properly using the definite description; what he actually asserted, if anything, would have to be determined by investigation of his intentions. If there is such an object, then it is the referent of the definite description; the speaker's *assertion* is true or false depending on whether the object is or is not brown. But his *statement* is false, since it presupposes that no other object is a table in the room. The third major difficulty concerns the fact that the speaker asserted *of whoever will be the first U.S. president of the 21st century* that he will be a Democrat. If we say the speaker *referred to* that president, then if that president turns out to be Gordon Flash, we will have to say that the speaker referred to Mr. Flash and asserted of him that will be a Democrat. Nevertheless, provided we make sure we are not taken to say that the speaker *stated* that Mr. Flash will be a Democrat, there appears to be nothing false in the contention that the speaker referred to Mr. Flash.

Indeed, given that that president will be Mr. Flash, the truth or falsity of the speaker's assertion depends only on whether Mr. Flash will be a Democrat.

9. As we have explicated *reference*, a speaker *refers to* the individuals in the state of affairs corresponding to the proposition he asserts. We can say that the *referent* of a given term or phrase in the sentence used by the speaker is the individual whose being referred to is accomplished by the use of that term or phrase. If a speaker asserts truly that there is one and only one table in his room, then he does not refer to the table since the relevant state of affairs, the "indefinite" state consisting of there being one and only one table in his room, does not consist in *that table's* being the only one in the room. In contrast, when a speaker asserts truly *of the table in his room* that it is brown, then the state of affairs corresponding to the proposition asserted is the state consisting of *that table's* being brown, and hence he is said to *refer to* the table. If the assertion was accomplished in the statement "The table is brown," then the table is the referent of the definite description 'the table.' And if the assertion was accomplished in the statement "The desk is brown," then the table is the referent of the definite description 'the desk'; in this case, the speaker's *assertion* is true while his *statement* is false, since he implies that the table is a desk and the only desk in the room.

10. In using a definite description the speaker *presupposes* that the referent *fits the description*, i.e., is *characterized* by the *concept* associated with the description. For example, the sentence 'The table is brown' has a negate 'The table is not brown'; by a use of either sentence a speaker would imply that the referent is such that (1) it virtually possesses the relevant attribute, say, of being a table in his room, and (2) no other object is a table in his room. If the referent happens to be the only table in his room, then the statements made would have true presuppositions. If the referent exists but is not a table, then presupposition (1) is false, since to say something virtually possesses an attribute is to say if it exists it does possess the attribute. If there are other tables in the room, presupposition (2) is false. If the referent does not exist, then (1) is true, for it follows that either the referent does not

exist or it is a table in the room, and hence that if it does exist then it is a table in the room. (This follows because we are here using 'if ... then ...' in the "weak" sense of material implication.) Thus if the referent does not exist, then it virtually possesses the relevant attribute. If the referent does not exist and if there are no other things which are tables in the room, then both presuppositions are true. If the referent does not exist, then if there is some other object which is the only table in the room, then even though the definite description *denotes* this object, the speaker has not *referred to* this object. Hence, even though it is true that this object is the only table in the room, and even if this object is brown, the speaker's statement is false due to the falsity of one of his presuppositions, i.e., that there were no objects *other than the referent* which were tables in the room. (If the referent does not exist, and there are no tables in the room, then any other non-existent object x will be such that (1) x is virtually a table in the room, and (2) no other objects are tables in the room. But it does not follow that the speaker *presupposes* that these things are true of x. The speaker presupposes that such things are true of *the referent*, and by hypothesis x is not the referent.)

11. In using a definite description, a speaker not only presupposes that the referent fits the description, but also *creates the presumption* that the referent exists. A speaker in uttering a sentence *creates the presumption* that p if and only if the proposition p will become a presupposition of the speaker's unless the proposition is *cancelled*.[1] A speaker in uttering a sentence *cancels* a proposition if and only if the speaker implies that the proposition is or may be false; in other words, if and only if there are at least some features of the speech situation such that proper use of the sentence in the situation implies that the speaker does not or may not believe the proposition. For example, Speaker D above, who made the two statements "The present king of France is not bald. He doesn't exist" used the definite description 'the present king of France' to refer to the present king of France, the same individual referred to by Speakers A and B. Conse-

[1] Cf. Grice, "The Causal Theory of Perception," *op. cit.*, 446.

quently D presupposed that the king was virtually king of France and that no one else was king of France. In using this description D also *created the presumption* that the king existed, as did A and B. But in D's case, unlike the others, the proposition that the king existed was *cancelled*, for there were some features of the speech situation, specifically his next statement, which were such that proper use of the sentence 'The present king of France is not bald' in this situation implied that D did not believe that the king existed. Thus, while the proposition that the king existed, being in the cases of A and B a *non-cancelled presumption*, constituted a presupposition of A's and of B's, in the case of D the proposition was a *cancelled presumption*, and hence D cannot be said to have presupposed that the king existed. In general, we can say that since a speaker using a definite description creates the presumption that the referent exists, *normally* a speaker using a definite description presupposes that the referent exists.

In ordinary discourse there are many "cues" (especially those associated with fiction, myth, legend, make-believe, fable, and dream reports) which signal that the presumption of existence associated with definite descriptions is being *suspended*. In such cases, and also in cases in which the existence of the relevant object is the question at issue, *assertions* that the relevant objects have certain attributes are to be interpreted as assertions of the *virtual* possession of the attributes. Thus there is no need to claim that such discourse is not referential, or that some kind of "secondary" or "fictional" reference is taking place, or that there is a special kind of existence called *fictional* existence. Since the presumption of existence has been suspended in such discourse, the statements involved constitute developments, elaborations extensions, creations or clarifications of the *concepts* of the individuals involved. In the case of certain complex forms of discourse involving discussions concerning whether a certain fictional or legendary character is identical with a certain real individual, there is a subtle mixture of ordinary statements and statements with suspended presumptions of existence. In investigations directed toward determining whether or not a certain object or person exists, it is particularly important that there be some means of repeatedly referring to that individual and of continuously clarifying and elaborating the concept of the individual, for it may require the prolonged efforts of many investigators to collect enough true information about the individual to settle the question of its existence.

12. If the statement "The table is brown" and the statement "The first U.S. president of the 21st century will be a Democrat" are both construed as involving referential uses of the relevant definite descriptions, how can we account for the obvious difference between the two, the difference expressed by saying that in the latter, unlike in the former, the speaker asserts of *whoever* will be the first 21st century US president that he will be a Democrat?[1] To account for this difference we do not have to postulate that in the latter statement the definite description is being used non-referentially. The relevant definite descriptions are being used in the same way in both cases, i.e., to refer to the relevant individuals; and each description is associated with the presupposition that the individuals virtually possess the relevant attributes and that no other individuals possess them, i.e., with the presuppositions that the individuals fit the descriptions. But there are features of the speech situation (the subject-matter in this case) which indicate that in the latter statement, the speaker *in addition* presupposes that the referent *by definition* fits the description, while in the former statement the speaker *in addition* presupposes that the referent merely *happens* to fit the description. While the definite descriptions are being used in the same way, the *statements*, due to additional contextual factors, have *more specific* presuppositions which differ in kind: while the first statement contains the presuppositions that the referent (that table) is such that (1) it is *synthetically* true that it is virtually a table in the room, and (2) it is *synthetically* true that nothing else is a table in the room, the second statement contains the presuppositions that the referent (that president) is such that (1) it is *analytically* true that he is virtually a 21st century US

[1] For a full discussion, see Donnellan, *op. cit.*

president, and (2) it is *analytically* true that there are no previous 21st century US presidents. In formal terms, an individual constant representing that president can be correctly introduced only subject to certain axioms or "meaning postulates" which place prior restrictions on what other individuals can be said to be identical with the individual referred to by the constant. For example, Gordon Flash cannot be said to be identical with that president unless Mr. Flash is virtually a 21st century US president and is such a president prior to everyone else. Of course, these propositions are not analytically true of Mr. Flash, even though he is identical with that president and they are analytically true of that president. It is a *synthetically* true fact that Mr. Flash has all of the attributes of that president; that the relevant propositions are analytically true of that president is not something which is an *attribute* of that president, and hence is not something which Mr. Flash must possess in order to be identical with that president. (The relevant propositions would be analytically true of Mr. Flash if it were *analytically* true that he was identical with that president, as would be the case if we were to introduce the term 'Mr. Flash' as a more convenient substitute for the longer term 'the first 21st century US president'.)

13. Thus, definite descriptions should not be said to be *ambiguous* – they have one and only one use, which is to refer to the relevant individuals. In a sense, a definite description could be said to be *vague* (indeterminate) with respect to the question concerning whether the referent fits the description *by definition* or *merely happens* to fit it. But it would be preferable to say that the *statement* may be vague (indeterminate) concerning this question. Indeed, in certain cases a statement may be said to be *ambiguous* concerning this question if the speech situation is one in which the statement must contain either the specific information that the referent fits the description by definition, or the specific information that it only happens to fit it, but where the audience is unable to determine which piece of information is contained in the statement.

With respect to statements which are fully determinate

concerning the referent's fitting the description by definition or by "accident," we can distinguish divergent properties in relation to the supposition that the presuppositions are false. Suppose a speaker makes the statement "The desk in Bertrand's office is red," and the context is such that the statement includes the presupposition that the object referred to *just happens* to fit the description. If there is an object (existent or non-existent) of which the speaker asserted that it was red, then the truth-value of the presuppositions is irrelevant to the truth value of this assertion. Since the presuppositions are non-analytic, they may be false; for example, perhaps the referent exists but is not a desk, or is not in Bertrand's office, or perhaps there is more than one desk in his office and the concept associated with the description is not specific enough to select one of them. If these presuppositions are false, the statement is false; moreover, it *may* (but need not) be difficult or impossible for the audience to know the *meaning of his utterance* (i.e., to know what the assertion was) because, knowing the *meaning of the sentence* used, the audience is apt to settle on some other proposition as the assertion. Nevertheless, there is no essential relationship between the truth-value of the presuppositions and the truth-value of the assertion.

Let us now suppose that the speaker makes the statement "The desk in Bertrand's office is red," where the context is such that the statement includes the presupposition that the object referred to fits the description *by definition.* (Perhaps the speaker is discussing Bertrand's fondness for red.) The speaker has asserted of *whatever* object is the desk in Bertrand's office that it is red; or perhaps we should say that this appears to be his assertion. If the speaker intended to assert of whatever object is the *chair* in Bertrand's office that it is red, then the latter is his assertion (and the audience may be able to determine from contextual indicators that the latter is his assertion). In such a case the speaker's *statement* is false because he presupposes that the referent (the chair) fits the description 'the desk in Bertrand's office', something which is false even though the presupposition that *the desk* fits the description is and must be true, since it is true by definition.[1] On the other hand, if the speaker intended to assert of whatever object is the desk in Bertrand's office that it is red, then if there is more than one desk in Bertrand's office, no assertion is made. Under such circumstances there cannot be any object (existent or non-existent) which is such that it is analytically true that every object fitting the description is identical with that object; by hypothesis, there are at least two *distinct* objects which fit the description. In formal terms, it is impossible under such circumstances to introduce an individual constant together with an axiom or "meaning postulate" to the effect that no other objects fit the description. Nevertheless, it would be misleading to say that in such a case the truth-value of the presuppositions affected the truth-value of the assertion; for this is a case in which no assertion and no presuppositions were actually made. (In many cases of descriptions in ordinary discourse, such empirical considerations do not bear upon the question of whether a genuine statement was made. For example, for our speaker's statement "The first US president of the 21st century will be a Democrat" there is a "logical" or "*a priori*" assurance of being a genuine statement, since by definition

[1] Donnellan recognized the possibility that even in "the attributive use" of a definite description, whatever is the object of which the assertion is made may fail to fit the description. (See "Putting Humpty Dumpty Together Again," *op. cit.*, 209.) But he did not seem to realize that in view of this possibility, all of the characteristics he lists as distinctive of "the referential use" hold also of the "attributive use." (See *ibid.*, 206.) To illustrate, in terms of the above example we can truly say:

(1) The speaker has referred to the chair even though it is in fact not a desk.

(2) He said something true or false about the chair even though it is in fact not a desk.

(3) He has, in using 'the desk in Bertrand's office' to refer to the chair, presupposed that the chair is a desk.

(4) In reporting the speaker's speech act, it is correct to say that he has stated something about the chair, and in reporting this we can use expressions (e.g., 'the chair in Bertrand's office') to refer to the chair other than the expression 'the desk in Bertrand's office' or synonyms of it.

(Of course, in relation to (4), it would be misleading to say that the speaker has *stated* that the chair in Bertrand's office is red.)

Donnellan, at various points in the articles cited, suggested that a Russellian analysis may be appropriate for "the attributive use" of definite descriptions. However, it seems scarcely less counterintuitive than in "the referential use" to say that a speaker who asserts of whatever is the chair in Bertrand's office that it is red has asserted that there is one and only one chair in Bertrand's office and that it is red.

not more than one can be first. If the US becomes a monarchy or adopts a dual presidency, then that president will simply fail to exist.)

14. Thus in any case of a genuine statement involving a definite description, the description has a single function – to refer to something – and the truth of the assertion made about that thing is not dependent on the truth-value of the presuppositions, nor upon their specific character as synthetic or analytic (if indeed the statement has this degree of specificity). When a speaker asserts something of the so-and-so, what he asserts is true (false) if and only if the so-and-so does (does not) have the attribute it is asserted to have. If there is nothing (existing or merely possible) of which the "assertion" is made, then no asserting has taken place; on the other hand, if there is such a thing (existing or merely possible), then even if the falsity of one or more of the presuppositions makes it difficult or impossible for the speaker to succeed in communicating with his audience, the truth value of his assertion depends only on whether that thing has the attributes he asserts that it has. Furthermore, if there is something (existing or merely possible) which is the referent of the definite description, then a genuine *statement* has been made, an entity which is well-defined conjunction of the propositions asserted and the propositions presupposed, and which always has a truth-value dependent solely on the truth-value of these propositions.[1]

15. If the theory outlined in this paper proves to be tenable, we will be able to maintain a "common sense" view of the role of definite descriptions in ordinary discourse, a view according to which definite descriptions always serve to refer to things (even to non-existent things, whose existence we should be able to deny). Moreover, the view will be associated with a well-known formal framework possessing an unusual combination of power and simplicity–standard quantification

[1] The difficult problems concerning definite descriptions in relation to identity statements, modal statements, and statements involving propositional attitudes are beyond the scope of the present treatment. Perhaps the notion of statement we have developed will throw light on these problems. For example, perhaps if propositional attitudes are explicated in terms of *statement* operators, substitutions of identical *statements* would not produce paradoxes.

theory. Finally, the explication of existence associated with the present theory of definite descriptions appears to enable us to make sense out of assertions and denials of existence without sacrificing our robust sense of reality.[1]

[1] I would like to express my appreciation to my colleagues Thomas Paxson, David Seligman and Robert Wolf, for helpful discussion and suggestions in relation to this paper.

DISTINGUISHING PRESUPPOSITION IN EPISTEMOLOGY

DONALD S. LEE
Tulane University

Presupposition is a method of inference used in finding or justifying an assumption underlying a given belief. As a method of inference it is not a matter of drawing the assumption from the given belief by necessity. Accordingly, I shall term presupposition to be a species (or "mode") of hypothetic inference[1] in that assumption is hypothetically asserted to be a logical condition for the given belief.

My central aim is to characterize presupposition as it is used in epistemological inquiry. To do so requires as much or more effort distinguishing it from other forms of reasoning as pointing out its own distinctive marks. But, the distinguishing is necessary if for no other reason than to overcome the doubts of those who cannot grant the existence of either presupposition or hypothetic inference.

The procedure used to accomplish my aim will be first to characterize and distinguish epistemology, second to characterize and distinguish presupposition, and finally to characterize and distinguish the use of presupposition in epistemology.

1. *Epistemology*

Philosophers have sometimes divided philosophy into three areas of interest: metaphysics, epistemology, and axiology.[2] I shall accept the three-fold division simply for the purpose of indicating the general character of epistemology.

[1] See my article, "Inferential Meaning in Philosophic Questions," *Tulane Studies in Philosophy*, XVII, (1968), 84–85. Hereafter this article is cited in parentheses in text as "*IM*."

[2] For instance: Everett Hall, *Philosophical Systems* (Chicago: University of Chicago Press, 1960), p. 3. He uses the term "ontological" instead of "metaphysical". Hereafter this work is cited in parentheses in test as "*PS*."

1 (i). In distinction from metaphysics and axiology, epistemology is a study in which the criteria of relaible knowledge are sought. In contrast to epistemology, metaphysics is concerned with laying bare the most general aspects (or categories) of being or with indicating the inner nature of reality; in another contrast to epistemology, axiology is concerned with uncovering the nature of value, with indicating the kind of thing value is as compared or related to fact or knowledge. Each area, epistemology, metaphysics, or axiology, as an intellectual enterprise seeks the fundamental principles of its dominant concern: epistemology seeks the fundamental principles of knowledge, metaphysics seeks the fundamental principles of being or reality, axiology seeks the fundamental principles of value.

1 (ii). The characterization of epistemology as being a study concerned with the principles of *knowledge* tends to mislead people into believing it is much narrower in scope than it has been since Kant. Some believe that since epistemology is concerned with knowledge, and since knowledge must be true, it follows that epistemology should essentially seek the criteria of truth. However, this view of what epistemology seeks is too narrow since the topic of truth is only part of the whole field of epistemology[1] ... unless one chooses to expand the meaning of the term "truth" to the point where it becomes synonymous with "knowledge".

A study of Kant shows that the problem of knowledge is directed, more or less generally, to the topic of meaning. For Kant, the forms of intuition and the categories were not addressed directly to the issue of the truth of judgments but more to the issue of how there could be any cognitive meaning whatever. After answering that question, Kant could then raise the problem of the truth of metaphysical claims, where the problem is dealt with entirely within the strictures of meaningfullness. Currently, Everett Hall has characterized epistemology in terms of "reference" (*PS*, 3). This characterization illustrates that epistemology is not confined to the

[1] Truth is a semantic property; semantics is a part of semiotics; and semiotics is a topic in epistemology.

quest for the criteria of truth, but investigates the wider topic of meaning.

To depict epistemology as primarily concerned with meaning rather than with truth leads into the puzzle of saying that one is searching for the principles of knowledge and yet is not directly aiming for criteria of truth. The puzzle arises from the common-sense doctrine which claims that since one cannot know anything unless it is true, it follows that truth must be at the core of knowledge. One way to unscramble the puzzle is to consider a probabilistic theory of empirical knowledge which defines truth in an all-or-nothing sense. Truth in this case would be a regulative idea. Any empirical claim would be more or less probably true. In any given instance our reports on and claims about cognitive experience would end in probablity, and the non-experienced ideal of truth would function as the definition by means of which probability would make sense. Simply because a knowledge claim fell short of the ideal defined by truth, there would be no reason to say that the claim was not knowledge. Thus, actual knowledge would be held in terms of probability, and epistemology would extend beyond the issue of truth criteria at the same time that it utilized those criteria for a standard which knowledge approximates.

From the brief considerations above we can see that it is not necessary for knowledge to be absolutely true nor for epistemology to be concerned principally with the search for truth criteria. We have seen one particular way in which epistemology is not limited to the topic of truth, which indicates it can be concerned with the wider topic of meaning or reference. From the time of Kant this latter topic, meaning, probably has occupied center-ring in the epistemological arena more often than any other topic.

2. *Presupposition in philosophy*

Epistemology has just been briefly delineated in order that I may ultimately show how presupposition operates within it. Now there remains a need for several comments on how presupposition operates in all of philosophy, particularly in two other fields of philosophy; to wit, metaphysics and axiology.

"Presupposition" is the specific name for assumption-seeking when used in philosophy. (Historically speaking, we can recognize that Kant and Collingwood both used the term in a technical sense, each meaning something somewhat different but nevertheless similar in many respects. They both contended that the main thrust of philosophic reasoning proceeds by presupposition, though each claimed that it had some characteristics which do not accord with what I have said and have yet to say.) In general, the method proceeds by first accepting as credible a general belief about some kind of phenomenon or other, and then searching to find principle(s) that could constitute grounds for the belief. The search must be a stepwise procedure; the search is not for a set of ultimate categories to posit first and then develop and elaborate until arriving at the original general belief. The search is for a *proximate* ground underlying the belief; it is not a matter of making large jumps to remote grounds, jumps which would skim over a train of thought needing more thorough step-wise connections.

2 (i) Presupposition can be more finely characterized in terms of the kind of starting point from which the whole inferential series begins and the kind of goal towards which the whole series proceeds. Very generally speaking, it begins with ordinarily accepted beliefs, or perhaps just one belief, and proceeds toward some very general criteria or principles. (Whether the beginning beliefs have true reference to things beyond themselves is irrelevant to their status as starting points.) As one works through presuppositions underlying presuppositions, etc., one successively lays bare implicit conditions behind implicit conditions. At what depth in the levels of presupposition one chooses to call the implicit conditions by the name "principles" is mainly a verbal issue, but only at the deepest levels are the conditions general enough to be called "principles" instructively. The process of successively revealing implicit conditions leads to the deepest principles, and those principles make it clear why the beginning belief from which the process started is believable anp

why it is worthy of belief. An example of a beginning point might be the belief that I am at present sitting at my desk writing. An example of a principle reached might be what Mill called the uniformity of nature.

2 (ii). It will be helpful for purposes of contrast to see very briefly how presupposition might be used in metaphysics and axiology before seeing its use in epistemology. In metaphysics, presupposition would begin with an ordinary belief about everyday objects or phenomena and try to uncover step-by-step that which is real as distinct from that which is apparent and that which is unreal in the original belief. The ultimate goal would be to reach the fundamental principles of reality; e.g., God or matter. Along the way, the levels of presupposition would gradually approach closer to fundamental principles. In axiology, presupposition would start with the report of a specific kind(s) of ordinary value experience and would gradually move from it, step-by-step, to general underlying principles of value. Such an investigation could be termed "inquiry into valuation". In such inquiry the question is not whether specific value experiences are valuable; rather, they are accepted as such while the inquiry is directed to uncovering successively the general principles involved in the original value situation. An example of a beginning point might be the "intuition" (immediate feeling) of rightness in some situation. An example of a fundamental principle ultimately presupposed, and revealed at the end of the inquiry, might be freedom.

2 (iii). In this brief section I delineated presupposition in philosophy, pointing out its general beginning point and general goal therein. I then mentioned in a few phrases how it operates in metaphysics and in axiology. These specific operations contrast in terms of beginning points and goals with its specific operation in epistemology.

3. *Presupposition in epistemology*

Epistemology is a study of the criteria of reliable knowledge; its ultimate aim is to find the fundamental criteria (often called "principles" at very deep levels) underlying the reliability of any belief claimed to be knowledge. When epistemology is undertaken by searching for presuppositions, it is not a matter of first positing criteria at the deepest level and then proceeding from there to a description of already existing knowledge. Rather, it is a matter of beginning with beliefs already accepted by virtue of some extra-philosophic reasons such as scientific reasons, commonsense reasons, or authoritarian reasons, for example. As a particular example of a beginning point in epistemology there are the forms of judgments behind which Kant sought criteria; i.e., conditions for the possibility of the judgments being knowledge. Epistemological beginning points are beliefs which lie "close to the surface" of experience. (the philosopher's experience is probably more enriched conceptually than that of the ordinary man.) The philosopher infers by presuppositional inference from the beliefs to criteria with which he classifies those beliefs and judges them to be meaningful and true. Ultimately, he aims for non-necessary conditions which can ground the possibility of there being any such thing as knowledge.

3 (i). Right off, it may be wondered how to justify one proposed beginning point over another or others. It is not that only one beginning point is appropriate in any presuppositional inquiry in epistemology; but, there must be ways to distinguish an appropriate one from inappropriate ones. It is justifying this distinction which is at issue here. Such justification can be given in two complementary ways: first, in terms of the striving for system; and second, in terms of the "naturalness" of a proposed beginning point.

3 (i) (a). First, if one aims to sytematize any body of knowledge, then an epistemological justification of the system's beginning points involves considering the subject matter at issue. When any given subject matter provides the literal or explicit meaning of a formal system it is an interpretation of that system. Such a system is built upon primitive terms which combine to form axioms. Whenever any primitive term is theoretical, i.e. not correspondent with observable experience, it has empirical meaning only "implicitly".[1] For instance, "electron" is justified as "implicitly" meaningful by

[1] R. B. Braithwaite, *Scientific Explanation* (Cambridge: Cambridge University Press, 1953), pp. 76–87.

virtue of the successful application of the atomic theory to the subject matter of sub-microscopic entities. In general, we can speak not only of the *meanings* of theoretical primitives being "implicit" but also of the *justifiability* of a set of primitives being "implicit"; namely, in terms of their ability to generate a formal system which successfully covers some given subject matter. The successful covering of some given subject matter by a formal system can be determined according to four criteria: 1) the logical interrelatedness of the system, 2) the width of scope of the original subject matter being covered, 3) the fittingness of distinctions generated or constructable within the formal system onto the original subject matter, and 4) the fruitfulness of the system in anticipating heretofore unsuspected relationships which when brought to attention are seen to be so. In short, part of the general justification of primitives as knowable concepts lies in the success with which the system in which they are primitive covers the subject matter about which knowledge is desired.

3 (i) (b). The second, and more specific, justification of a beginning point is directed to its "naturalness". The naturalness of a beginning point is a function of the meaning it has before being considered as the beginning point of a system. In philosophic explanation, as distinct from scientific explanation, some glimmering of the meaning of a primitive must be somewhat familiar and vaguely understood by those for whom the explanation is intended. For instance, Kant used the idea of "judgment" (proposition, statement) as a primitive from which he reasoned by presupposition, and Descartes used the ideas of "thinking" and "being". These concepts were philosophically significant and already had existing beliefs about them *before* those thinkers spelled out their systems. (To be sure, their systems charged those concepts with much new meaning that fed back and changed the old meaning.) In general, some previous communicability of meaning is essential for the beginning points of any philosophic field.

The pre-systematic communicability of primitive explanatory terms is set by cultural usage and conceptualization in the manner indicated by Collingwood who maintained that metaphysical significance is historically dependent on a conceptualization of the world prevalent during a given epoch.[1] Somewhat similar to this is Nelson Goodman's notion of "entrenched" predicates, a notion which holds that systematic conceptualization of the experienced world depends on ideas which are already the unquestioned currency of previous modes of conceptualization.[2] Two tests of the "naturalness" of a primitive (not mentioned by Collingwood or Goodman) are: 1) how often specified concepts have been used in the past as explanatory terms, and 2) how many people, among the philosophically enlightened or perhaps among all the learned people of a given historical epoch, seriously use the concepts in question. If both investigations would reveal that a concept is used extensively and unquestioningly then it would be a "natural" primitive ... assuming, of course, that a systematic field could be built using it and some other primitives.

3 (ii). When using presupposition one starts with ordinary experience and looks for principles behind it. If the kind of ordinary experience is perspicuous, a beginning point can be easily grasped. Now, language can be noticeably perspicuous in a *prima facie* way, and epistemology accepts linguistic expressions as its beginnings, hence epistemology can have perspicuous beginning points. In other words, what we call reliable knowledge comes in the form of linguistic expressions, so by taking such expressions as beginning points of epistemological inquiry we are not immediately thrust into the problem of looking beyond them to other philosophic dimensions such as "reality" or "value". We can set out upon our inquiry, so to speak, from the *prima facie* form confronting us. Language captures concepts and statements symbolically and makes them directly accessible for investigation.

In contrast to epistemology by presupposition, metaphysics

[1] R. Collingwood, *An Essay on Metaphysics* (London: Oxford University Press, 1962), pp. 49–77.

[2] Nelson Goodman, *Fact, Fiction and Forecast* (New York: Bobbs-Merrill, Inc., 1965), pp. 87–99.

by presupposition refers to the nature of the world rather than to the nature of judgments about the world; thus, it has a more difficult task in characterizing its beginning points. Such metaphysical beginning points of presupposition are the ordinary objects in the ordinary world. Unfortunately for metaphysics by presupposition, what is meant by "ordinary objects in the ordinary world" is problematic, it is not perspicuous in a *prima facie* way. In short, the subject matter of epistemology falls much more in line with the method of presupposition than does metaphysics, since the former in dealing with language expressions has perspicuous beginning points while the latter does not.

3 (iii). When inferring by presupposition in epistemology it is tempting to look for existential conditions temporally preceding a given situation rather than to look for presuppositions. Such existential conditions are often and easily mistaken for presuppositions.

3 (iii) (a). The mistaking of temporally antecedent conditions for presuppositions is made not only at the particular level of presupposition but also at the more general level of assumption-seeking. For instance, if one were seeking to find what making stew assumed he might mistakenly conclude that since meat, potatoes, and carrots were conditions for stew they were "assumptions" underlying stew. But they are not; they are existential conditions . . . particular ingredients in this case. Or perhaps one might think fire is an assumption underlying the making of stew. Again, this is a mistake. Though fire is not an ingredient like meat, it is also not an assumption; rather, it is a causal condition needed to make the stew be what it is. An assumption behind a situation is not merely a temporally precedent condition. An assumption is a logical condition, and though it is sometimes extensionally identical with an existential condition it is not the same intensionally. Accordingly, the concept "making stew" assumes what it is to be an ingredient without getting involved in whether any ingredients exist or what they are. Furthermore, the concept of making stew assumes blending tastes irrespective of the means for bringing the blending about, whether by fire or by mystical incantations. The concept of

making stew as it is ordinarily understood assumes the concepts of ingredient and blending though it is not necessarily defined as such.

3 (iii) (b). The above example concerned assumption-seeking in general, but, presupposition as a particular philosophical kind of assumption-seeking is more directly our interest here. Presupposition in epistemology is a matter of seeking a logical condition underlying a possibility of a given epistemological concept. Thus, in full, a presupposition of X is a logical condition underlying a possibility of the concept of X, where X is a term of general epistemological interest. To confuse such presuppositional inference with the tracing of existential conditions (ingredients or causes) is to mistake it (presupposition) for another mode of hypothetic inference called "successive-complexing" (*IM*, 87–88). Though the latter is not an incorrect form of hypothetic inference in itself, it is not proper to confuse it with presupposition.

3 (iii) (c). Some epistemologies have been worked out rather roughly by successive-complexing for instance, those of Locke and Dewey. Let's examine these briefly for the purpose of contrasting this mode of inference so often confused with presupposition.

3 (iii) (c) (1). In the first example, Locke supposes that a man's mind is a blank slate at birth. Then simple ideas are impressed through the senses onto the slate which accumulates more and more of them. Ideas from inner feelings of the mind's operations also arise to join those other ideas already recorded on the slate. Then combinings take place forming various kinds of abstract, compound, complex, and relational ideas. Then complex and simple ideas give rise to general ideas. For Locke, that which can be understood about the nature of knowledge are the general conditions under which it can be held by the mind. Then he gives us an understanding of such general conditions by setting down one condition and then successively adding more conditions, as if they occur by such a temporally cumulative process in the mind. As Locke completes his explanation he makes it successively more complex.

3 (iii) (c) (2). As a second example, Dewey presents an

epistemological explanation of inquiry in terms of the inferential mode I have called successive-complexing. The general background for his explanation of inquiry is his belief that man as a cognitive creature has a natural tendency to recognize and to try to solve problems in order to reach a state of equilibrium with his environment. The original state is one of equilibrium, a state which is "blank" relative to the events that call forth the thinking process. The entrance of disturbing factors is the first stage, and the recognition of the disturbance as a problem upsetting equilibrium is the second stage, and so on until the problem is resolved. The final result is that equilibrium returns on a more complex level. In the process of solving the problem, criteria are developed, criteria guiding the return to equilibrium. Each succeeding stage of the process can be conceived as existing by virtue of the existence of a preceding stage which provided the existential conditions from which the succeeding stage emerged.

3 (iii) (c) (3). Neither Locke nor Dewey constructed their epistemologies by presupposition. Each did it by successive-complexing. However, it is very tempting to think that such tracing of successive existential conditions reveals the presuppositions underlying knowledge, especially when one is not aware of the genuine difference between the two modes of hypothetic inference involved, presupposition and successive-complexing.

3 (iii) (d). In order to point up this difference between the two modes, let us look at an example in epistemology contrasting inference by presupposition to inference by successive-complexing.

Consider Kant's attempt to explain the plausibility of the categories of the understanding. From one point of view (successive-complexing) he says that because sense-intuition provides the content unified by the categories, those categories depend on sense intuition. The inference from sense-intuition to categories is one of successive complexing because there is a progression from the concept of one level of conditions to the concept of a higher level of conditions which, though the higher unify the lower, depend on the lower for their very existence. On the other hand, when Kant explains the plausibility of the categories from the presuppositional point of view, he starts from ordinary linguistic judgments, abstracts their form, and then claims that the judgment-forms presuppose the categories. With this type of inference (presupposition) he arrives at the categories as criteria for the application of ordinary concepts to the sense manifold, where such application occasions our knowledge of the manifold. The inference from forms of judgment to categories is by presupposition because there is a progression from judgments as given (about sense experience) to the categories as their underlying logical conditions.

3 (iv). A somewhat different distinction concerning presupposition must be noted here. In characterizing presupposition as the search for "a logical condition underlying the possibility of the concept of X" I did *not* say or mean "a logical condition underlying the possibility of *conceiving* X". The latter would refer to logical conditions underlying the psychologically characterized states of mind one would tend to have or need in order to conceive X. Such logical conditions would be the parameters in, or the formal conditions of, a psychological theory. Such scientific matters are not our concern herein.

The distinction between an analysis of psychological conditions involved in conceiving and an inference by presupposition can be clarified by an example, again referring to Kant's epistemology. Consider an explanation of Kant's categories of the understanding by a theory of psychology: such a theory would be set up to explain how the mind could actually operate so that there could be categories which are non-empty and unifying. The logical conditions underlying such a theory could, mistakenly, be thought to be presupposed by the categories of the understanding.[1] Suppose there were a psychological theory of conceiving modeled upon Kant's "blue glasses" analogy in which the understanding would be considered to be a kind of seeing or opening of the mental "organ"

[1] Formally, a presupposition is similar to an explanatory premise in a scientific theory. But scientific theoretic premises lack the semantic (meaning) and pragmatic (historical) aspects which tie presuppositions into language as ordinarily used.

to some kind of mind-stimuli (not sense stimuli) from outside that organ. The organ would have a structure which would determine what, of all that comes in, would register and the form in which it would be registered. The organ would be a scanner of its environment for input. Obviously, in this psychological theory, conceiving (understanding) would be modeled upon perceiving. Further along these lines the theory would depict mental stimuli being emitted by some kind of "objects" to which the understanding would have to adjust for the purposes of mental equilibrium.

If the above would be the rough outline of a psychological theory then the logical conditions underlying that theory would be its parameters or its formal characteristics. The parameters in this case would be organ, stimuli, mental form, scanner, mental equilibrium, while the more formal conditions of the theory would be the logic of the language associated with biological, physiological, and perhaps cybernetic features of the theory. Except for "mental form" perhaps, none of these conditions are directly pertinent to *presuppositions*, properly understood, underlying the categories of the understanding. Even "mental form" as utilized in the psychological theory is not directly relevant to the epistemological context concerning presupposition since in the psychological theory the concept of category (mental form) is viewed entirely in terms of its psychological operation.

Contrasted to logical conditions underlying a psychological theory of the categories are logical conditions underlying the concept of the categories; the latter logical conditions are presuppositions. There are three candidates in Kant's epistemology for such a role and either or all may be presuppositions. Listed in order of the likelihood of their being presuppositions are: the transcendental unity of apperception, the Ideas of Reason, and the schematism. There is no need to proceed further with Kant here since my aim has been only to indicate a possible confusion that may arise between presuppositions and parameters or formal conditions of a psychological theory of conceiving. By differentiating between organ, stimuli, register, scanner, etc. on the one hand and transcendental unity, the Ideas, and the schematism on the other hand, the example shows that presuppositions are not parameters of a psychological theory.

3 (v). In this section I concentrated on the investigation of presupposition in epistemology. The investigation inquired into the nature of the beginning points and the nature of the inferential moves involved therein. The kind of beginning point was indicated to be some belief accepted as credible for extra-philosophic reasons. As to whether or not a beginning point for a philosophic system is a "good" one depends on two considerations: first, implicitness ... namely, the successful covering of some given subject matter by the system generated from the beginning point; and second, naturalness ... namely, the pre-systematic usage and communicability of the terms used as the beginning point. After treating the topic of the beginning point I then considered presupposition as an inferential move in epistemology. The move is facilitated by inquiring systematically into the meanings of linguistic expressions, because in epistemology we seek criteria of knowledge where such knowledge is expressed in the form of language.

In order to clarify how a presuppositional inference is made I contrasted it with other moves often confused with presupposition. Two confusions were noted and illustrated. The first confuses presupposition with the search for existential condiitons temporally preceding a given situation. Such a confusion has at least two varieties: mistaking ingredients as presuppositions and mistaking physical causes as presuppositions. This first confusion mistakes another inferential mode, successive-complexing, with presupposition. The correct characterization of presupposition stands in contrast to this kind of confusion; properly speaking, presupposition is the search for a logical condition underlying the possibility of a given concept. The second confusion about presuppositional inferences in epistemology lies in mistaking it to be a search for logical conditions underlying the possibility of *conceiving* something rather than of the *concept* of that something. The former calls for a search for either the parameters or the formal conditions of a psychological theory of conceiving; the latter calls for a search for underlying meanings on which a given conception is based.

Though other confusions concerning the proper identification and use of presupposition undoubtedly exist, these two, i.e. the confusion between presupposition and successive-complexing and the confusion between conditions of a psychological theory of conceiving and those of the meaning of a conception, are the most subtle, prevalent and misleading ones.

4. *Summary*

In concluding I will highlight the essential points covered herein in order to present in one overall picture the general nature of presupposition in epistemology. First I differentiated epistemology from some other fields of philosophy; namely metaphysics and axiology. Epistemology is the search for criteria of reliable knowledge. It is not so narrow as to be concerned solely with truth, but is wide enough to be concerned with meaning. I then delineated presupposition in philosophy to be the seeking of assumptions behind a given belief. I mentioned the possibility of using presupposition in metaphysics and axiology only for providing a contrast to its use in epistemology.

The methodology was elaborated more fully when I explained how presuppositional inference is used in epistemology. In systematic epistemology the beginning point of inference by presupposition is some epistemological belief accepted for extra-philosophic reasons, and its adequacy is measured in terms of fruitfulness of development and in terms of its naturalness. Then the inferential move by presupposition was defined as the search, by consideration of meanings involved in a concept, for a logical condition underlying it. Such logical conditions are not to be confused either with existential conditions or with conditions underlying a psychological theory of conceiving.

5. *Concluding remark*

It is, indeed, only too easy to confuse presupposition with other modes and types of reasoning. But, presupposition is not alone in this predicament of being easily confused with other reasoning. For example, it is very easy in a conversational context to confuse deductive or inductive inference with analogical or genetic reasoning. Only reflection by analysis will reveal the outline of what is involved in a given instance of reasoning.

As reasoning animals, we naturally utilize various types of inference. We do not start reasoning in a specific way because some philosophers distinguished it from others. But, we can reason *better* in a specific way if philosophers recognize and distinguish that way from other ways and weed out fallacies. And, to satisfy our philosophical yearnings, we all desire to reason better.

VERIFICATION IN METAPHYSICS

PETER M. BURKHOLDER
Central Washington State College

Harold N. Lee has contended that metaphysical doctrines, such as solipsism and mind-matter dualism, should be interpreted as hypotheses.[1] This claim is of considerable importance, for if it is correct, we should reject not only the traditional dogma that some metaphysical doctrines are "absolutely certain," but also the skeptical rejoinder which the dogma frequently has evoked, namely that they cannot be knowledge at all. Lee's proposal is a mean between these two extremes, since it permits metaphysical doctrines to be probable knowledge. It has the advantages of each extreme, and the defects of neither.

Although Lee intends his view to apply to all metaphysical doctrines whatsoever, he does not present it categorically. On the contrary, as he is careful to point out, it is itself an hypothesis and therefore subject to testing and possible disconfirmation (*MH*, 352). He admits, of course, that some of the very greatest metaphysicians, such as Descartes, have not supposed their doctrines to be hypothetical, but he does not think this a serious objection against his view. For such doctrines very often have functioned as hypotheses in practice, no matter how their proponents may have understood them. Moreover, they can be entertained and evaluated as hypotheses by a present-day inquirer.

I find myself largely in agreement with this proposal. In the pages that follow, I shall explain it, with the help of some passages from Lee's writings, and develop my own version of it,

[1] He develops the view in "Metaphysics as Hypothesis," *Journal of Philosophy*, XLIV (1947), 344–352. Subsequent references to this article will be inserted in the text with the abbreviation "MH". Professor Lee's view is in the pragmatic tradition. It has family resemblances to theories proposed by C. S. Peirce, C. I. Lewis, A. N. Whitehead, and S. C. Pepper.

giving an account of the logical pattern that verification of a metaphysical hypothesis would display. Professor Lee very likely would not endorse all that is said in this paper. For the views expressed in it occasionally may diverge somewhat from his. However, I hope he would not strongly disapprove of the paper's general drift.

I

In order to clarify his proposal that metaphysics be considered hypothetical, Lee first discusses hypotheses in natural science. A scientific hypothesis, according to his account, is a "generalization or series of generalizations such that, if it is assumed, the particulars of experience will fall under it as special cases and thus be explained" (*MH*, 346). Any scientific hypothesis can be expressed as the antecedent of a conditional statement whose consequent contains a claim that, if certain conditions are satisfied, some specified body of data will become observable. The data are understood to be, or to have been generated by, particular instances of the generalization in the antecedent. For example, Newton's Law of Gravitation implies that after a rock has been hurled into the air, it will, under certain conditions, fall back to the earth. If the consequent of a conditional of this sort is true, the antecedent is probable to some degree; and if the consequent is false, the antecedent is false. Therefore, when a scientist wishes to test an hypothesis, he should, in effect, first formulate a conditional statement in accordance with this pattern, then conduct laboratory experiments, or otherwise collect observations, in order to determine, so far as needful, whether the consequent is true. If the hypothesis survives a thorough scrutiny of this kind, and no alternative hypotheses are found to be more probable, e.g. because of an ability to explain a wider range of data, the scientist may then regard it as knowledge, even though it is not fully certain.

An important (and only apparently paradoxical) implication of this account of the logic of science is that the very same data are subsumable, at least theoretically, under two or more alternative hypotheses, rendering each of them probable; and if all of them happened to be equally probable, each could be said to be knowledge. In practice, one would rarely encounter a situation of the latter sort. However, I suspect this is due merely to a lack of human ingenuity in framing fruitful hypotheses, and that for any body of data numerous alternative scientific hypotheses could serve with equal adequacy as an explanation.

After presenting his analysis of hypotheses in natural science, Lee develops his proposal that we should interpret metaphysical doctrines in a somewhat similar manner. They differ from the laws of science in that they are "the most general of all explanatory principles" (*MH*, 346). But since they are hypotheses, they should be understood, like scientific generalizations, as antecedents of conditional statements which one could formulate. Moreover, again like scientific hypotheses, their acceptability is dependent on the truth of their consequents. There are, however, some important differences between scientific and metaphysical hypotheses as regards verification: One verifies a scientific hypothesis by investigating whether its instances, as predicted in the consequent of the conditional, actually occur, or whether counter-instances occur instead. But one can not, in Lee's opinion, adequately verify a metaphysical hypothesis in this way. "Observation can not be as determinant for metaphysics ..." (*MH*, 347).

I believe Lee is correct that scientific and metaphysical verification differ. For, unlike a scientific hypothesis, once a metaphysical doctrine has been tentatively accepted, it comes to seem peculiarly *inevitable,* as if one had a right to expect, in advance of any empirical investigation, that nothing he could find in experience would be an ineluctable counter-instance to it; that, on the contrary, he would *always* be entitled to offer some explanation or series of explanations that would preserve it. For example, the hypothesis of the existence of other minds cannot be overturned by the observation of counter-instances, for any apparent counter-instance always may be explained away: "Those people are intoxicated," "They have a very low I.Q.," "They are out of their minds," etc. A scientific hypothesis, on the other hand, may *not* be preserved indefinitely by means of such a strategy, because the series of explanations would have to come to an end after a relatively short while in order for the hypothesis to be genuinely scientific. As Lee justly remarks, natural science must not "explain away the appearances ... It must fit its theoretic explanation to them, and if the attempted explanation does not fit, it, not the appearances, is at fault."[1]

The reason for this important difference between the two disciplines is that metaphysical principles, unlike the laws of natural science, are categorial *norms* by which we are to *interpret* experience. They are *definitive* of all things in experience, as regards their ultimate significance; and to accept such a principle is to acknowledge and acquiesce in the authority it has over them – something we can do even while feeling metaphysically nauseous. Metaphysics differs from natural science, therefore, in that all instances of its principles are shaped by the principles themselves. Moreover, if we were to accept a *consistent* set of such principles, we could admit no contrary particulars even as a possibility, for nothing we might envision as an object of experience could violate any of our principles. As Lee observes ,"in so far as the particulars of experience are distinct and formulated, they are already subsumed under categories" (*MH*, 351); i.e., they are instances of principles we already accept. And (in my opinion) this is true, not just of actual particulars, but of possible ones also.

It follows, from these same considerations, that if someone does try to condemn a metaphysical hypothesis, as he could a scientific generalization, on the ground that things in his experience are counter-instances to it, he is simply indicating that it may be logically inconsistent with some of his own principles. In itself, this would by no means constitute a disconfirmation. It would not even show that the rejected hypothesis is wrong for *him*, because he himself very well may need to revise his categorial commitments, his ways of interpreting and dividing up the world, and alter his view as to what things he could encounter in experience. For example, if a philosopher rejects the principle of universal determinism

[1] "Scientific Method and Knowledge," *Philosophy of Science*, X (1943), 69.

on the ground that introspection reveals uncaused acts of will, it may indeed be the case, but would not necessarily *follow,* that the determinist hypothesis is false. For what may be the case instead is that the philosopher should surrender whatever metaphysical principle is leading him to interpret his experience non-deterministically, and adopt the determinist view, thereby committing himself to give a determinist interpretation of everything. Moreover, it would not be easy to ascertain which of these two alternatives would be correct. The very existence of the alternatives shows, however, that metaphysics differs from natural science.

In light of the foregoing discussion, there would be little point in trying to verify a metaphysical hypothesis precisely on the scientific model. Therefore, Lee is correct that "the details of method in the ... verifying of hypotheses will differ in the two fields" (*MH*, 347). Accordingly, if we propose to treat metaphysical doctrines as hypotheses, we should be prepared to explain, at least in outline, how to verify them in some other way than by observing whether any counter-instances occur. We should be able to give an account of an alternative mode of verification that will preserve their hypothetical character and yet do justice to the peculiar features that differentiate them from hypotheses in natural science. In the next section of the paper, I shall first give an interpretation of Lee's views regarding this matter, and then present some of my own.

II

"It would be the height of presumption on my part," Lee states, "to say that I had worked out the methods of verification of metaphysical hypotheses."[1] Nevertheless, he does offer some proposals as to how metaphysical verification should proceed. For example, "the only choice" between alternative metaphysical hypotheses "rests upon the degree to which they ... make experience intelligible – the degree to

1 "Reply to Mr. Berenda's Comments on 'Metaphysics as Hypothesis'," *Journal of Philosophy*, XLV (1948), 301. Needless to say, I do not presume to claim that I have worked them out, either.

which they enable us to respond successfully to the given element in experience and to understand" (*MH*, 351). A verified metaphysical hypothesis is "able to order experience so that we can 'get around in it' both by action and by understanding" (*MH*, 347). It enables us to "make judgments of real and unreal that will not bring us either to grief in our practical affairs or to frustration in our intellectual enterprises" (*MH*, 347–8). I think we would not seriously misrepresent the purport of these passages if we concluded that, according to them, we may verify a metaphysical hypothesis by ascertaining that, as a result of the categorizing activity which it permits, it possesses both *practical value* and *instrumental cognitive value* for those who accept it, and none of its available alternatives has more. A brief (but harmless) way of expressing this criterion is to say that a metaphysical doctrine is probable knowledge if and only if it *maximizes* both practical value and instrumental cognitive value.

A verified metaphysical hypothesis has practical value, according to my understanding of Lee's criterion, in that if one used it, either expressly or tacitly, to interpret his experience, the interpretation would tend to assist him in his goal-seeking behavior. Some metaphysical systems promote free-flowing, effectual action, which enables one to attain his ends with relative ease, and such systems therefore have much practical value. However, others are inhibiting, demoralizing, and debilitating, and therefore unsatisfactory from a practical point of view. One's ends, of course, inevitably are influenced by whatever system he happens to accept. For instance, a typical Baptist's ends differ from those of a typical Buddhist, in part as a result of their metaphysical differences. Therefore, if we were to evaluate several alternative hypotheses to determine which is most suitable for a given person, we might have to judge each hypothesis in relation to a somewhat different set of ends. But we could still compare the alternatives as regards their degree of satisfactoriness. We might find, for example, that hypothesis H would be highly satisfactory in serving ends R, which some specific person would have if he accepted H, but that hypothesis J would do very little to serve the somewhat different ends S, which would arise for that person if he accepted J. In that case, H would have more practical value for him than would J. And if no available alternative to H should be more satisfactory in helping him to achieve his ends, including whatever ends he might acquire as a result of accepting that alternative, then H would maximize practical value for him.

A verified metaphysical hypothesis also has instrumental cognitive value, in that, if accepted, it would help one to attain understanding in the various non-metaphysical areas of inquiry, such as history, biology, physics, mathematics, aesthetics, ethics, and politics. It would supply a framework of concepts for interpreting the subject-matter of these disciplines; and the interpretation would tend to assist one in making successful predictions, cogent explanations, or valid deductions within each of them. Since, as Lee points out, "understanding probably must be defined in terms of the categorial activity" (*MH*, 347), what constitutes understanding very likely would differ somewhat from system to system. Nevertheless, a metaphysical hypothesis could be judged as regards its ability to promote its own version of understanding. For instance, if hypothesis H is highly satisfactory in promoting understanding, as defined in its own terms, while hypothesis J does very little to promote understanding, as defined in *its* terms, H has more instrumental cognitive value than does J. And if no available alternative to H is more satisfactory in promoting understanding, as defined in the alternative's terms, then H maximizes instrumental cognitive value.

Lee gives some examples of verified metaphysical hypotheses, including Descartes' dualism of mind and matter. During the seventeenth century, at any rate, this doctrine was exceedingly helpful to the physical scientists who accepted it. They used its categories "to order the phenomena with which they dealt. It enabled them to 'find their way around' in their field of investigation" (*MH*, 349). At present, however, Cartesian dualism seems to have outlived its usefulness. As Lee observes, "Most of us ... think today that the reasons for rejecting it are sufficient" (*MH*, 349). This suggests that verification of a metaphysical hypothesis is relative to cultural situations. Indeed, since a doctrine which maximizes practical or cognitive value for one group of persons may fail to do so for another group, it can be probable in relation to some contexts but improbable in relation to others. Therefore, if someone who accepts Lee's conception of metaphysics, or a similar one, should refer to a doctrine as knowledge, he would not necessarily be endorsing it for every situation whatsoever, nor would he imply that he believes it himself, or thinks highly of it. For example, we could legitimately claim that the ideology of a Hitler or a Caligula is knowledge if, in addition to promoting nonmetaphysical understanding, it would enable him to attain his goals. But this in no way indicates that we would approve of him or his goals, nor of his beliefs, except *solely* in relation to his own cultural situation.

There is a further consequence of the view that metaphysical verification is relative, namely that when one evaluates a doctrine to determine whether it is appropriate for a specific cultural situation, he should take full account of the various relevant characteristics, including purposes and existing beliefs and practices, of the persons actually in the situation. One's interpretation of an alien situation inevitably will be influenced by his own prior categorial commitments, but he should nevertheless study it as conscientiously and as respectfully as possible. Such a policy would increase the probability that if his own situation should become the object of such an investigation, he would be treated with equal consideration.

In some of Lee's later writings, evidently so that his view can be distinguished more sharply from sentimentalist and voluntarist forms of pragmatism, practical value is deemphasized as a criterion of verification in metaphysics, while instrumental cognitive value continues to be stressed.[1] Some philosophers of intellectualist predilections might wish to carry this approach even further and make instrumental cognitive value the sole criterion of verification in metaphysics, so that a metaphysical hypothesis would be considered verified if and only if it maximized instrumental cognitive value, whether or not it maximized practical value also. On the

[1] Cf. "Philosophy and the Categories of Experience," *Tulane Studies in Philosophy*, VII (1958), 69, 71, *et al.*

other hand, other philosophers, such as the sentimentalists and voluntarists, might prefer to make practical value the sole criterion. A decision in favor of instrumental cognitive value would not, of course, necessarily result in practical value being ignored altogether, for things that have instrumental cognitive value often, for that very reason, have some practical value too. For instance, when understanding is itself an end, as it occasionally is, anything that assists us in attaining it thereby acquires not only instrumental cognitive value but also practical value. Moreover, even when understanding is only a means to other ends, what produces it can still be said to have practical value indirectly. Therefore, a metaphysical doctrine that has instrumental cognitive value could come, in either of these ways, to have practical value as well. Indeed, in such situations, though in no others, instrumental cognitive value would be a *variety* of practical value. This is not the whole story, however, for there are numerous human ends; and since conflicts can arise among the ends, in that the attainment of one can prevent the attainment of another, something that has practical value in relation to one end can have practical dis-value in relation to others. Therefore, even if a doctrine had instrumental cognitive value of the practical variety, it could also have so much practical dis-value of other sorts that its over-all practical value, taking into account the various purposes it would serve or thwart, would be negative.

Although we can evaluate a metaphysical doctrine as regards its instrumental cognitive value, we can also do so as regards its other values, including (other) specific forms of practical value, and over-all practical value. These are distinct matters, even though they overlap in various ways; and when one speaks of the "verification" of a doctrine, he should be prepared to specify what kind or kinds of value he has in mind. If, as may be best, we should decide to treat instrumental cognitive value as the sole criterion of metaphysical verification, we must then face the crucial issue of whether it would be better, in that case, to recommend to someone a "verified" doctrine or one that has other values instead. And we might decide that an "unverified" one sometimes would be preferable. In my opinion, whatever criterion we employ in metaphysical verification, we should endeavor to see, in determining whether a doctrine is appropriate for someone, that his *various* interests, not just those promoted by instrumental cognitive value, are equitably served by the choice.[1] If that means we must recommend an "unverified" one for him, so be it! However, I strongly suspect that, for most persons, a doctrine which maximized instrumental cognitive value would maximize overall practical value also. Admittedly, this view, too, is an hypothesis, but it is an *a posteriori*, not an *a priori*, one.

Accordingly, the logical pattern of verification for metaphysical doctrines, which allows them to be hypothetical even though they depart somewhat from the scientific model, requires that if we are to verify such a doctrine, we should make it the antecedent of a conditional statement whose consequent contains a claim that the doctrine, if accepted by someone, would maximize some specified value or values for him. If we find the consequent to be true, we are entitled to infer that the antecedent, which contains the doctrine, is probable knowledge in relation to his situation. However, if we find that some alternative hypothesis is more satisfactory for him, and that the consequent therefore is false, we should infer that the antecedent also is false and that he should reject the doctrine.

This analysis of metaphysical verification allows for the possibility that two or more alternative metaphysical hypotheses may be equally satisfactory for someone, and thus equally probable. Therefore, several competing systems all may be knowledge, provided that no alternative to them is more satisfactory. This is the metaphysical analogue of an important feature of scientific hypotheses which we noted earlier, namely that alternative hypotheses can explain the same data. And it indicates that pluralism of metaphysical belief may be entirely justifiable, not because each system could explain the same data – that possibility is ruled out,

[1] In this process, one may criticize and alter some of the interests themselves, as John Dewey would suggest.

since each would generate its own data – but because each could maximize the same value or values.

III

Sometimes Lee advocates an additional test for adequacy in a metaphysical hypothesis, namely logical consistency (*MH*, 347). It pertains not just to a single doctrine that is proposed for acceptance, but to the entire system of doctrines that would be accepted at the same time. One question that arises, in connection with this test, is how logical consistency is related to maximization of practical value or of instrumental cognitive value. Is it a sufficient condition for either of those criteria, a necessary condition, both, or neither? It is probably not sufficient, for a consistent system can have considerable dis-value, of either the practical or the cognitive sort. Moreover, although I personally abhor logical inconsistency, its absence does not seem to me to be a necessary condition for maximizing either of those values. For, in the first place, perhaps *no* available metaphysical system is logically consistent; and, in the second place, the interests of some people, such as the mystics, may be served better by inconsistency than by consistency.

One might dispute my conclusion that logical consistency is not necessary for maximizing either practical value or instrumental cognitive value, on the ground that since a logical inconsistency implies any statement whatsoever, a person who accepted such an inconsistency would thereby commit himself to numerous propositions to which he would not really wish to be committed, and this would produce not only much cognitive dis-value but also considerable practical dis-value, since it would frustrate so many of his purposes. However, such an argument would no longer be so plausible as it once would have been. In a recent article, the logicians Alan Ross Anderson and Nuel Belnap showed that not all statements can legitimately be inferred from a contradiction, because some inferences of that sort would commit a fallacy of *relevance*. And they presented rules for determining when an inference from inconsistent premises would be improper.[1] As a result, contradictions can be *contained;* though often inconvenient, they are not necessarily always fatal. It is true they would be very detrimental in a system intended for a cultural situation in which logical consistency is a strongly desired end-in-itself. Some of the great Continental Rationalists are examples of persons in such a situation. But even for thinkers so clear-minded as they, some logically inconsistent system could very well be more satisfactory than any available consistent one, if we take into account their other interests and purposes.

Since logical consistency is therefore neither necessary nor sufficient to maximize practical value or instrumental cognitive value, we may infer that, if a metaphysical doctrine is adequate if and only if it does maximize one and/or the other of these values, logical consistency is neither necessary nor sufficient for metaphysical adequacy. However, it is nevertheless highly relevant to the value of a metaphysical hypothesis. In many but not all types of cultural situation, logical consistency serves as a very helpful, but not essential, *means* to both practical value and instrumental cognitive value. Without it, one often would encounter more difficulty in ordering experience than he would if his system were closely similar but logically consistent. The *ease* which logical consistency produces does indeed contribute to a system's utility. However, the contribution would not always be large enough to insure that some available alternative system, that happened to be inconsistent, could not have greater utility as a result of certain advantages that would over-compensate for its inconsistency. For example, some forms of solipsism probably are logically consistent, while Cartesian dualism is not. Yet, in spite of this, the latter is far more useful.

A related difficulty, namely pragmatic inconsistency, sometimes arises as a result of accepting a metaphysical doctrine. It may be defined, rather broadly, as any conflict

[1] "Tautological Entailments," *Philosophical Studies*, XIII (1962), 9–24. A logician may invent a system in which contradictions are omni-implicative. But such a system should be carefully distinguished from any system in which anyone ever has seriously formulated his beliefs.

that occurs within a person's actions, purposes, and beliefs, and is not reducible to a logical inconsistency. It can take various forms. It occurs, for instance, when a person adopts a policy that thwarts his dominant purposes, or when his several ends conflict with each other, or when he commits himself to a set of ideals or social conventions but fails to live up to them, or when he does not treat like cases alike. Very often, philosophers are pragmatically inconsistent. For example, as Berkeley wryly noted, many of his contemporaries declared secondary qualities to be subjective, on the ground that they are relative to an observer, but inconsistently refrained from drawing the same conclusion about primary qualities, even though they too are relative to an observer. Another case of pragmatic inconsistency is that of the legendary solipsist who wished to be praised by the public.

Pragmatic inconsistency is a defect. Therefore, if someone adopts a metaphysical doctrine that produces such an inconsistency for himself, he is subject to criticism. Moreover, the doctrine also very likely should be criticized – at any rate, in relation to that person's situation. For pragmatic inconsistency is a source of both practical and cognitive disvalue.

IV

In this paper, we have examined the logical pattern of verification in metaphysics, using as a starting point Lee's proposal that metaphysical doctrines be understood as hypotheses. As the argument developed, we confirmed his observation that scientific and metaphysical verification differ, explaining how and why they do so. Next, we indicated some of the alternative forms that metaphysical verification might take. Finally, we commented, not unfavorably, on Lee's view that logical consistency is a virtue in metaphysics, and then suggested an additional (but related) virtue, namely pragmatic consistency. The conception of metaphysics that Lee advocates has thus turned out to be promising. What we need now is a much more detailed and highly specific account of the several ways in which metaphysical hypotheses may help to advance understanding and facilitate practice.

GENTZEN'S CUT ELIMINATION THEOREM FOR NON-LOGICIANS

LARRY W. MILLER
Tulane University

The most important theorem of constructive mathematical logic, Gentzen's cut elimination theorem, is largely unknown even among those acquainted with mathematical logic. Usual treatments of mathematical logic, oriented to semantics, use formalization only to find a syntactic property (theoremhood) to correspond to the semantic property (validity). What is needed for a more meaningful use of formalization (and so of mathematical logic) is an effective way of studying formal proofs. The importance of Gentzen's cut elimination theorem lies in the fact that it gives a normal form for formal proofs, so they can be looked at meaningfully. I.e., one can read off certain results about a theorem simply by looking at its normal form proof.[1] The purpose of this paper is to present Gentzen's cut elimination theorem for mathematically minded philosophers.[2] Since our interest is in understanding, we shall give up the usual rigor of logical treatments of this theorem and present it in the spirit of mathematics.

If logic is the study of valid inference among statements, then to make this study mathematical (and so get mathematical logic) we must idealize the inferences and statements so they can be treated mathematically. So we must formalize. Natural language becomes a precisely defined formal language

[1] To show a formal system consistent, it suffices to show that f is not a theorem where f is a false statement. It usually will not be evident that there is no proof of f, but we may be able to see directly that there is no normal form proof of f.

[2] To understand everything perhaps one needs a bit of mathematical sophistication, but the main drift should be clear in any case. Some of the tricks of exposition which in my opinion make this treatment of Gentzen's cut elimination theorem more readable than usual treatments were inspired in part by W. W. Tait's treatment of infinitary propositional calculus. See his "Normal derivability in classical logic" in *The Syntax and Semantics of Infinitary Languages*, Springer Lecture Notes in Mathematics, vol. 72 (1968), 204–236.

– nouns and sentences respectively become terms and well formed formulas (wffs), etc. Some of the wffs are logically simple (do not contain any logical symbols (other than =)) – we call these atoms. Other wffs (molecules?) are built up from atoms by negation, conjunction, disjunction, implication, equivalence, universal quantification, and existential quantification – we use $\sim$, $\wedge$, $\vee$, $\supset$, $\equiv$, $\forall$, $\exists$ for these respectively.

In usual treatments of mathematical logic (after one has specified what it means for a wff to be true in an interpretation of the formal system), one syntactically singles out certain wffs (the theorems or provable wffs) by giving axioms and rules of inference. The set of theorems is the least set of wffs containing the axioms and closed under the rules of inference. The idea is to make the set of theorems equal the set of wffs which are valid (true in all interpretations).
E.g., one can take as axioms:

$$A \supset . B \supset A$$
$$A \supset (B \supset C) . \supset . (A \supset B) \supset (A \supset C)$$
$$\sim\sim A \supset A$$
$$\forall x A(x) \supset A(t)$$
$$\forall x(A \supset B) \supset . A \supset \forall x B \quad \text{if x not free in A}$$
$$t = t$$
$$t = t' \supset . A \equiv A' \quad \text{where A}' \text{ results from A by replacing some occurrences of t in A by t}'.$$

For the logician. In the fourth axiom t must be free for x in A(x). In the last axiom t must not occur in A in the scope of a quantifier on a variable free in t or in t′.
One can take as rules of inference:

$$\frac{A \quad A \supset B}{B} \quad \text{(modus ponens)}$$

$$\frac{A}{\forall x A} \quad \text{(generalization)}$$

Here $\dfrac{A_1 \ldots A_n}{B}$ means if $A_1, \ldots, A_n$ are all theorems so is B.
This axiom system is designed to make the Deduction Theorem (which makes theorem proving easy) easy to prove. We

call this axiomatization of first order logic with identity L. L_i will be L minus the identity axioms and L_q will be L minus the quantifier axioms and the generalization rule (so L_i and L_q axiomatize first order logic and propositional calculus with free variables and identity axioms, i.e., elementary calculus). We write $\vdash A$, $\vdash_i A$, or $\vdash_q A$ if A is a theorem of L, L_i, or L_q. By Gödel's Completeness Theorem $\vdash A$ if and only if A is true in all normal interpretations (where $=$ is interpreted as identity).

As we have mentioned, to make more use of a formal system than just to match the syntactic property of theoremhood with the semantic property of (normal) validity – Gödel's Completeness Theorem – what one needs is a way to study formal proofs effectively. Ideally, one needs to be able to find a simplest proof or a proof in some normal form for any theorem. To facilitate the search for a normal form for proofs, we reformulate our formal system L so we derive not wffs but finite lists of wffs. In this way and by making axioms trivial and putting the meat of the system in the rules of inference, we can isolate the role of each logical connective. We call the lists of wffs *sequents* and denote them by capital Greek letters. (Semantically, one can think of the sequent $A_1, \ldots, A_n$ as $A_1 \vee \ldots \vee A_n$, i.e., the sequent $A_1, \ldots, A_n$ is to be true if A_i is true for some $i = 1, \ldots, n$.)

Sequent calculus

Axioms: $A, \sim A$ where A is an atom
$t = t$ where t is a term

Rules of inference:

$$\frac{\Gamma, A \qquad \Gamma, B}{\Gamma, A \wedge B} \qquad (\wedge\text{-rule})$$

$$\frac{\Gamma, A, B}{\Gamma, A \vee B} \qquad (\vee\text{-rule})$$

$$\frac{\Gamma, A}{\Gamma, \forall x A} \quad x \text{ not in } \Gamma \qquad (\forall\text{-rule})$$

$$\frac{\Gamma, A(t)}{\Gamma, \exists x A(x)} \quad t \text{ free for } x \text{ in } A(x) \qquad (\exists\text{-rule})$$

$$\frac{\Gamma, A \qquad \Gamma, \sim A}{\Gamma} \qquad \text{(cut-rule)}$$

$$\frac{\Gamma}{\Delta} \quad \text{if each wff in } \Gamma \text{ occurs in } \Delta \qquad \text{(S-rule)}$$

E.g. of S-rule:

$$\frac{A, B}{A, A, B}, \quad \frac{A, A, B}{A, B}, \quad \frac{A, B}{B, A}, \quad \frac{A, B}{A, B, C}, \text{ and } \frac{A, B}{A, B}.$$

Given the S-rule our $\wedge$- and cut-rules are equivalent to

$$\frac{\Gamma, A \qquad \Delta, B}{\Gamma, \Delta, A \wedge B} \text{ and } \frac{\Gamma, A \qquad \Delta, \sim A}{\Gamma, \Delta}.$$

We have one rule for each logical connective except negation. What about negation and the second identity axiom for L? We find it convenient to assume that for each atom A there is an atom $\sim A$ such that $\sim\sim A = A$. We take as further syntactic identities:

$\sim(A \vee B) = \sim A \wedge \sim B$
$\sim(A \wedge B) = \sim A \vee \sim B$
$\sim\forall x A = \exists x \sim A$
$\sim\exists x A = \forall x \sim A$
$A = A'$ if A' results from A by replacing some occurrences of a term t in A by a term t' where $t = t'$ (with the previously mentioned restriction).

$\Vdash \Gamma$ means Γ is a theorem of sequent calculus. $\Vdash^{cf} \Gamma$ means Γ is a theorem of sequent calculus without use of the cut-rule. We can now state Gentzen's cut elimination theorem: If $\Vdash \Gamma$ then $\Vdash^{cf} \Gamma$, i.e., the cut-rule is superfluous and can be eliminated.

LEMMA For all wffs A we have $\Vdash^{cf} A, \sim A$.

PROOF By induction on the logical complexity of A. If A is an atom then A, $\sim A$ is an axiom. If $A = B \vee C$ then by the induction hypothesis $\Vdash^{cf} B, \sim B$ and $\Vdash^{cf} C, \sim C$ so we can put

$$\dfrac{\dfrac{\begin{array}{c}|\\ B, \sim B\end{array} \qquad \begin{array}{c}|\\ C, \sim C\end{array}}{B, C, \sim B \wedge \sim C}\wedge}{B \vee C, \sim(B \vee C)}\vee$$

The proof for $A = B \wedge C$ is dual to this (Interchange wffs with their negations). If $A = \forall xB$ then by the induction hypothesis $\Vdash^{cf} B, \sim B$ so we can put

$$\dfrac{\dfrac{\begin{array}{c}\vdots\\ B, \sim B\end{array}}{B, \exists x \sim B}\,\exists}{\forall xB, \sim\forall xB}\,\forall$$

The proof for $A = \exists xB$ is dual to this.

THEOREM $\Vdash A_1, \ldots, A_n$ if and only if $\vdash A_1 \vee \ldots \vee A_n$.
In particular, $\Vdash A$ if and only if $\vdash A$.

PROOF (only if) We must show $\vdash A \vee \sim A$ and $\vdash t = t$ and that the theorems of L are closed under (translations of) the rules of inference of the sequent calculus. E.g., (for the $\wedge$-rule) if D is a disjunction of wffs of L then we must have if $\vdash D \vee A$ and $\vdash D \vee B$ then $\vdash D \vee (A \wedge B)$. All this is well known.

(if) We must show the translations of the axioms of L are theorems of sequent calculus and that the theorems of sequent calculus are closed under (translations of) modus ponens and generalization. The latter is trivial since the translations are the cut- and $\forall$-rules.

Axiom 1 translates to $\sim A, \sim B, A$ which is a theorem of the sequent calculus by the Lemma.

Axiom 2 translates to $A \wedge B \wedge \sim C, A \wedge \sim B, \sim A, C$ and we have by the Lemma

$$\dfrac{\dfrac{\dfrac{\begin{array}{c}\vdots\\ A, \sim A\end{array} \qquad \begin{array}{c}\vdots\\ B, \sim B\end{array}}{A \wedge B, \sim A, \sim B}\,\wedge \qquad \begin{array}{c}\vdots\\ C, \sim C\end{array}}{A \wedge B \wedge \sim C, \sim A, \sim B, C}\,\wedge \qquad \begin{array}{c}\vdots\\ A, \sim A\end{array}}{A \wedge B \wedge \sim C, A \wedge \sim B, \sim A, C}\,\wedge$$

Axiom 3 goes to $\sim A, A$. Axiom 4 goes to $\exists x \sim A(x), A(t)$ and we have

$$\dfrac{\sim A(t),\ A(t)}{\exists x \sim A(x),\ A(t)}\ \exists$$

Axiom 5 translates to $\exists x(A \wedge \sim B(x)), \sim A, \forall x B(x)$ and we have

$$\dfrac{\dfrac{\dfrac{A, \sim A \qquad \sim B(x), B(x)}{A \wedge \sim B(x), \sim A, B(x)}\ \wedge}{\exists x(A \wedge \sim B(x)), \sim A, B(x)}\ \exists}{\exists x(A \wedge \sim B(x)), \sim A, \forall x B(x)}\ \forall$$

The identity axioms are trivial so the theorem is proved.

Our reformulation of L as a sequent calculus is now justified. We observe that in the cut-rule A disappears so if we look at proofs using the cut-rule we do not know what wffs have been "cut" out of the proof. However, in each rule of inference other than the cut-rule, each wff that occurs in a premise also occurs (in some sense) in the conclusion. Hence, each wff in a cut-free proof of Γ occurs (in some sense) in Γ. We want to make this precise.

DEFINITION A is a subformula of A and of $\sim A$.
A and B are subformulas of $A \vee B$ and of $A \wedge B$.
For any term t A(t) is a subformula of $\forall x A(x)$ and of $\exists x A(x)$.
If A is a subformula of B and B a subformula of C then A is a subformula of C.
Nothing else is a subformula

Then each wff in a premise of any rule of inference other than the cut-rule is a subformula of a wff in the conclusion.

SUBFORMULA PROPERTY OF CUT-FREE PROOFS: Each wff in a cut-free proof of Γ is a subformula of a wff in Γ.

If we could prove Gentzen's cut elimination theorem (If $\Vdash \Gamma$ then $\Vdash^{cf} \Gamma$) then we would have a normal form for proofs which would be meaningful in the sense that no wff could occur which was not a subformula of a wff in the conclusion.

E.g. 1. If Gentzen's cut elimination theorem holds then

L, L_i, and L_q are consistent since f (where f is a false wff) cannot be a theorem of sequent calculus by the subformula property of cut-free proofs – f cannot have a cut-free proof.

2. Note that if Γ is =-free then any cut-free proof of Γ is =-free by the subformula property of cut-free proofs. Hence, if Gentzen's cut elimination theorem holds and A is =-free then $\vdash A$ if and only if $\vdash_i A$ (eliminability of =). In other words, L, first order logic with identity, is a conservative extension of L_i, first order logic.

3. Again note that if Γ is quantifier-free then any cut-free proof of Γ is quantifier-free by the subformula property of cut-free proofs. So if Gentzen's cut elimination theorem holds and A is quantifier-free then $\vdash A$ if and only if $\vdash_q A$. In other words, L is a conservative extension of L_q, propositional calculus with free variables and identity axioms. Exactly the same argument shows L_i is a conservative extension of propositional calculus with free variables.

REMARK Examples 2 and 3 show that with respect to =-free or quantifier-free wffs Gentzen's cut elimination theorem would enable one to show that = or quantifiers are ideal in the Kantian sense, i.e., their use (to prove =- or quantifier-free wffs) is non-essential. Hilbert wanted to justify the use of infinity in mathematics by showing it is ideal. Example 3 shows this exactly for infinity-as-quantifiers.

We want to turn now to the proof of Gentzen's cut elimination theorem. But first we need a result which says roughly that the ∧-, ∨-, ∀-, and ∃-rules can be inverted; i.e., given a (cut-free) proof of the conclusion, one gets a (cut-free) proof of each premise.

INVERTIBILITY LEMMA

(1) If $\Vdash^{cf} \Gamma, A \wedge B$ then $\Vdash^{cf} \Gamma, A$ and $\Vdash^{cf} \Gamma, B$.

(2) If $\Vdash^{cf} \Gamma, A \vee B$ then $\Vdash^{cf} \Gamma, A, B$.

(3) If $\Vdash^{cf} \Gamma, \forall x A(x)$ then for any term t $\Vdash^{cf} \Gamma, A(t)$.

(4) If $\Vdash^{cf} \Gamma, \exists x A(x)$ then $\Vdash^{cf} \Gamma, A(t_1), \ldots, A(t_n)$ where $t_1, \ldots, t_n$ are all the terms such that the ∃-rule is used in the cut-free proof of Γ, ∃xA(x) in the form

$$\frac{\Delta_i, A(t_i)}{\Delta_i, \exists x A(x)}\ \exists.$$

PROOF Replace each occurrence of $\left|\begin{array}{c} A \wedge B \\ A \vee B \\ \forall x A(x) \\ \exists x A(x) \end{array}\right|$ which leads to the occurrence of $\left|\begin{array}{c} A \wedge B \\ A \vee B \\ \forall x A(x) \\ \exists x A(x) \end{array}\right|$ in the conclusion by $\left|\begin{array}{l} A \text{ or by } B \\ A, B \\ A(t) \\ A(t_1), \ldots, A(t_n) \end{array}\right|$ to get a cut-free proof of $\left|\begin{array}{l} \Gamma, A \text{ or of } \Gamma, B \\ \Gamma, A, B \\ \Gamma, A(t) \\ \Gamma, A(t_1), \ldots, A(t_n) \end{array}\right|$. (In the case of (3) one must further substitute t for x in the proper places.) Each (affected) instance of a $\left|\begin{array}{c} \wedge \\ \vee \\ \forall \\ \exists \end{array}\right|$-rule becomes an instance of the S-rule so the proof is complete.

GENTZEN'S CUT ELIMINATION THEOREM If $\Vdash \Gamma$ then $\Vdash^{cf} \Gamma$.

PROOF Since there are only finitely many cuts in any proof it suffices to show

If $\Vdash^{cf} \Delta, A$ and $\Vdash^{cf} \Delta, \sim A$ then $\Vdash^{cf} \Delta$.

This enables us to eliminate one cut at a time until none are left. We show this by induction on the logical complexity of A. In other words, the procedure is:

1. We show how to replace any cut on a logically complex wff by at most a finite number of cuts (all on logically simpler wffs).
2. We show how to eliminate any cut on an atom (a logically simple wff).

Case 1. $A = B \wedge C$. By the Invertibility Lemma $\Vdash^{cf} \Delta, B$ and $\Vdash^{cf} \Delta, C$ and $\Vdash^{cf} \Delta, \sim B, \sim C$ so we can put

$$\dfrac{\begin{array}{c} | \\ \Delta, B \end{array} \qquad \dfrac{\begin{array}{c} | \\ \Delta, C \end{array} \qquad \begin{array}{c} | \\ \Delta, \sim B, \sim C \end{array}}{\Delta, \sim B}\ \text{cut on C}}{\Delta}\ \text{cut on B}$$

We have replaced one cut on $B \wedge C$ by two cuts on B and C which by the induction hypothesis can be eliminated.

Case 2. $A = B \vee C$. Dual to Case 1.

Case 3. $A = \exists x B(x)$. By the Invertibility Lemma $\Vdash^{cf} \Delta, B(t_1), \ldots, B(t_n)$ and $\Vdash^{cf} \Delta, \sim B(t_i)$ for each $i = 1, \ldots, n$ so we can put

$$\dfrac{\dfrac{\Delta, \sim B(t_n) \qquad \Delta, B(t_1), \ldots, B(t_n)}{\Delta, B(t_1), \ldots, B(t_{n-1})}\ \text{cut on } B(t_n) \qquad \Delta, \sim B(t_{n-1})}{\Delta, B(t_1), \ldots, B(t_{n-2})}\ \text{cut on } B(t_{n-1})$$

$$\vdots$$

$$\dfrac{\Delta, \sim B(t_1) \qquad \overline{\Delta, B(t_1)}}{\Delta}\ \text{cut on } B(t_1)$$

We have replaced one cut on $\exists x B(x)$ by n cuts on $B(t_i)$ for $i = 1, \ldots, n$ each of which can be eliminated by the induction hypothesis.

Case 4. $A = \forall x B(x)$. Dual to Case 3.

Case 5. A is an atom. Take the cut-free proof of Δ, A and replace each occurrence of A which leads to the occurrence of A in the conclusion by Δ. The result is a cut-free proof of Δ, Δ (and so by the S-rule of Δ) unless $A, \sim A$ was used as an axiom. After the replacement this becomes $\Delta, \sim A$ but we have a cut-free proof of $\Delta, \sim A$ so we tack it on wherever needed and we're done.

COROLLARY 1 L, L_i, and L_q are consistent.

COROLLARY 2 If A is $=$-free then $\vdash A$ if and only if $\vdash_i A$. So L is a conservative extension of L_i.

PROOF If $\vdash A$ then $\Vdash^{cf} A$ and since A is $=$-free by the subformula property of cut-free proofs, the identity axioms are never used.

COROLLARY 3 If A is quantifier-free then $\vdash A$ if and only if $\vdash_q A$.
So L is a conservative extension of L_q.

PROOF If $\vdash A$ then $\Vdash^{cf} A$ and since A is quantifier-free the quantifier axioms are never used.

Before we state the next Corollary, a generalization of Corollary 3, we must explain what the Herbrand resolution of a wff is. With each wff A in L we associate a prenex normal form[1] wff H(A) – the Herbrand resolution of A – in an extension of L such that $\forall$ does not occur in H(A).

DEFINITION If the prenex normal form of A has the form $\exists x_1 \ldots \exists x_n \forall y B(x_1, \ldots, x_n, y)$ where $B(x_1, \ldots, x_n, y)$ is in prenex normal form then $H(A) = H(\exists x_1 \ldots \exists x_n B(x_1, \ldots, x_n, f(x_1, \ldots, x_n)))$ where f is a new n-ary function constant. We have eliminated one $\forall$-quantifier so if we repeat this procedure we get our Herbrand resolution of A.

E.g. If $A = \forall x \exists y \forall z B(x, y, z)$ then $H(A) = \exists y B(c, y, f(y))$ where c is a new individual constant and f is a new 1-ary function constant.

Since $\vdash \exists x_1 \ldots \exists x_n \forall y B(x_1, \ldots, x_n, y)$ if and only if $\exists x_1 \ldots \exists x_n B(x_1, \ldots, x_n, f(x_1, \ldots, x_n))$ is a theorem of L with the new function constant added, it is easy to show

> A is a theorem of L if and only if H(A) is a theorem of L* where L* is L plus all new function constants used to form H(A)

COROLLARY 4 (FIRST ε-THEOREM) If A is in prenex normal form and $\vdash A$ then $\vdash_q A_1 \vee \ldots \vee A_n$ where each A_i is a substitution instance of the matrix of A.

PROOF Since A is a theorem of L H(A) is a theorem of L* and so $\Vdash^{cf} H(A)$. By repeated use of the Invertibility Lemma (4) $\Vdash^{cf} A_1, \ldots, A_n$ which is quantifier-free so $\vdash_q A_1 \vee \ldots \vee A_n$.

The First ε-Theorem is a key theorem for proof theory of first order logic. Its importance lies in the facts that (1) it enables one to prove Herbrand's Theorem which gives a constructive necessary and sufficient condition for theoremhood (in L), (2) it enables one to prove easily Gentzen's erweitete Hauptsatz which enables one to permute inferences, (3) it underlies Hilbert's method of proving the consistency of various formalized theories in a constructive manner (see

[1] A wff A of L is in prenex normal form if it is of the form $Q_1x_1 \ldots Q_nx_nB$ where each Q_i is either $\forall$ or $\exists$ and B, the matrix of A, is quantifier-free.

the Consistency Theorem), and (4) it justifies the non-essential use of quantifiers to prove quantifier-free wffs.

REMARK If we would have proved a cut elimination theorem for ε-calculus we could have given a very easy proof of the Second ε-theorem which says ε-calculus is a conservative extension of L.[1]

We turn now to Herbrand's Theorem. The degree of a term t is one greater than the maximum degree of any proper subterm of t (so if t has no proper subterms it is of degree one). If A is in prenex normal let A* be the set of new function constants used to form H(A) plus an individual constant (if not already present). For $0 < m$ the *m-reduction* of A is the disjunction of all wffs got by substituting for the free variables in the matrix of H(A) terms which are of degree $\leq m$ and which are built up from symbols in A*.

HERBRAND'S THEOREM If A is in prenex normal form and contains no free variables then $\vdash A$ if and only if for some m the m-reduction of A is a tautology.

PROOF (only if) If $\vdash A$ then by the First ε-Theorem $\vdash_q A_1 \vee \ldots \vee A_n$ where each A_i is a substitution instance of the matrix of A so $A_1 \vee \ldots \vee A_n$ is a tautology. Replace any terms not made up solely of symbols in A* by an individual constant in A*. If m is the maximum degree of any term in the result then we have a subdisjunction of the m-reduction of A so the latter is a tautology.

(if) If the m-reduction $A_1 \vee \ldots \vee A_p$ is a tautology then $\vdash_q A_1 \vee \ldots \vee A_p$. For each $i = 1, \ldots, p$ we have a proof of H(A) in L* from A_i so H(A) is a theorem of L*. Hence, $\vdash A$.

We want now to give an application of the Herbrand Theorem to illustrate its usefulness. If A is of the form $\forall x_1 \ldots \forall x_m \exists y_1 \ldots \exists y_n B$ where B is quantifier-free and contains no n-ary function constants for $n > 0$ then it is decidable whether A is a theorem of L or not.

PROOF $H(A) = \exists y_1 \ldots \exists y_n B(c_1, \ldots, c_m, y_1, \ldots, y_n)$ where c_i is a new individual constant. The terms that can be built up from symbols in A* are $c_1, \ldots, c_m$ all of degree one.

[1] See A. C. Leisenring, *Mathematical Logic and Hilbert's ε-Symbol*, (London: McDonald & Co., 1969).

Hence, the l-reduction of A has mn disjuncts and so can be tested for tautologousness.

Finally, we mention a result used by Hilbert to prove constructively the consistency of various formal theories including a first order formalization of arithmetic restricted so that the induction axiom is quantifier-free.

CONSISTENCY THEOREM If Γ is a set of wffs of L and a valuation can be given which makes each substitution instance of the matrix of a wff in Γ true then Γ is consistent (constructively).

PROOF An easy generalization of the First ε-Theorem reads

If A and each wff in Γ is in prenex normal form and $\Gamma \vdash A$ then $\Delta \vdash_q A_1 \vee \ldots \vee A_n$ where each wff in Δ is a substitution instance of the matrix of a wff in Γ and each A_i is a substitution instance of the matrix of A.

Using this if $\Gamma \vdash f$ where f is some false wff then $\vdash_q \sim A_1 \vee \ldots \vee \sim A_n$ where each A_i is a substitution instance of the matrix of a wff in Γ. Hence, $\sim A_1 \vee \ldots \vee \sim A_n$ is a tautology so there is no valuation making each substitution of the matrix of each wff in Γ true.

THE AESTHETICS OF HAROLD N. LEE: A RECONSIDERATION

LOUISE N. ROBERTS
Newcomb College, Tulane University

As a student entering the field of philosophy, Harold N. Lee showed a major interest in aesthetics. *Perception and Aesthetic Value*, published originally in 1938 and reissued in 1967, is based upon his doctoral dissertation. If one considers his subsequent bibliography, however, it becomes apparent that this interest soon broadened and became increasingly diversified. But, a closer consideration of his work will reveal that, as his colleagues have suspected, his interest in aesthetics has never been lost and the problems of aesthetics have been a recurrent concern. Some of Lee's most recent writing has been in this area of philosophy.

It is not surprising that, despite the scope of his work, Lee has retained this interest. If one follows his aesthetics, one finds that, according to his analysis, aesthetic valuation lies at the core of human experience. Although he has illuminated issues that are relevant to the kaleidoscopic realm of art, he has offered something more basic than a philosophy of art in any restricted sense. His primary focus has been upon aesthetics in the etymologically traditional sense of concern with the perceptual. He locates aesthetic value in perceptual intuition, "... the direct awareness of that organization of data immediately apprehended through the senses or in sense imagery."[1] Aesthetic value is the value to be found within perception. It is intrinsic to perceptual apprehension. Although we in our sophisticated consciousness may only recognize such value through discursive analysis of myriad discursive forms (and it is to be remembered that all such distinctions are analytic)

[1] Harold N. Lee, *Perception and Aesthetic Value* (New York: Prentice Hall, 1938), p. 30. Hereafter references to this work will be inserted in parentheses in the text, symbolized by "*PAV*," followed by page number.

the initial perceptual grasp is immediate and not to be reduced to any discursive classification. The value is intrinsic to perception in that it lies within the perceptual grasp rather than in any of the conceptual forms which may contribute to perception or in any of the conceptual classifications to which a given perception may be subjected. As such its locus is essentially non-discursive. It is to be found in the immediate experiential forms which discourse is about. So it is that in aesthetic experience axiological and epistemological factors converge and this experience is as fundamental to man as perception itself. Aesthetics provides a keystone of humanistic philosophy in that it offers an interpretation of man's primary experience of value and an analysis of the foundations of human knowledge.

Although Lee's aesthetics can be related to theories of a number of other philosophers, such as Santayana, Croce, Bergson and D. W. Prall, his theory of perceptual intuition is derived from Kant. He has taken Kant to task, however, for changing the meaning of the term "aesthetic". In the *Critique of Pure Reason* Kant uses it to mean *Anschauung* (perceptual intuition). In the *Critique of Judgment* it is used to refer to the judgment of taste and means subjective rather than perceptual (*PAV*, 230). It is in connection with this criticism that Lee develops one of his most useful distinctions, the difference between valuation and evaluation.

Valuation refers to the apprehension of value. Evaluation refers to a judgment concerning this valuation. Lee has argued that there is a fundamental confusion lying at the base of Kant's theory, due to his method of approach. "It is the the confusion between the apprehension of beauty and the judgment concerning this apprehension; that is, the confusion between value and evaluation" (*PAV*, 237). Lee, like Kant, holds that the aesthetic experience is not a form of knowledge. It is noncognitive. According to Lee it is "feeling" in the sense of being a perceptual grasp of data immediately and directly present to the mind for which there may be no corresponding complete understanding (*PAV*, 45). No judgment is present. He points out that for Kant, "the apprehension of beauty is a result of the exercise of judgment.

Judgment is, for him, a cognitive faculty; that is, it is intellectual. Judgment is the subsumption of a particular under a universal" (*PAV*, 233). Lee grants that Kant attempts to avoid this implicit intellectualism by "recourse to the 'reflective judgment' or the subsumption of a particular which is given under a universal which is not given (that is, of which we are not aware when we are making the judgment)" (*PAV*, 234). The judgmental form remains, however, and it is this form which, when applied to value, Lee would distinguish from perceptual intuition, i.e. the apprehension of value, and would regard as *evaluation*. Evaluation is cognitive. It would appear that, in Kantian terms, evaluation would involve the subsumption of a particular under a universal. This requires an activity of the understanding.

Such activities are clarified by Lee's distinction between perceptual intuition and "perception of fact" which is synthetic and discursive. He remarks that perceptual intuitions are "what is given to consciousness, and it is by means of them that one makes judgments of fact, but when one is making the judgment of fact he is not regarding the data in their own nature and for their own sake" (*PAV*, 30). The colored shapes, sounds, etc. of perceptual intuition are regarded as signs to be subjected to factual classification. So it is that perception of fact contains an aesthetic element although it is not as such an aesthetic activity. As was remarked above, aesthetic valuation lies at the core of human experience, including the cognitive experience of the understanding as found in perception of fact.

It would appear that evaluation occurs at a rather high cognitive level. As has been indicated, according to Lee to evaluate is to make a judgment concerning value. He remarks that the most usual occasion for making value judgments is the occasion for comparing values. This would require recognition of the values to be compared and a basis for comparison. Lee finds the basis for comparison to be pleasure which he describes as the "evaluating factor" for aesthetic value. This "factor" introduces a hedonistic element into Lee's aesthetics.

It does not follow, however, that this theory can be dismissed as simply another statement of aesthetic hedonism.

Lee defines aesthetics at the philosophical study of value that is to be found in the data of perceptual intuition (*PAV*, 98). His consideration of pleasure is merely one factor in this study; moreover, aesthetic pleasure is pleasure of a particular sort. Aesthetic pleasure means perceptual pleasure. For one thing it is not to be identified with the pleasures of mere sensation. Perception is distinguished from sensation in that it involves form while sensation is amorphous. According to Lee, "aesthetic value, in the broadest sense, is a property attributed to an object by virtue of the fact that it can be perceptually apprehended with pleasure or displeasure" (*PAV*, 88). "Pleasure" is a synonym of pleasantness and pleasantness is described as "that affective state of consciousness, the conditions of which tend to be repeated or held or kept or continued only for the sake of being in that state" (*PAV*, 76). Unpleasantness, on the other hand, is "that affective state of consciousness the conditions of which tend to be avoided, not for any conscious ulterior reason, but only for the sake of not being in that state" (*PAV*, 76). Aesthetics, the philosophical study of value that is to be found in the data of perception, is not restricted to the pleasant but would include situations that involve the unpleasant as well. As Lee expresses it, value is polar, values are immediately apprehended as positive or negative (*PAV*, 68).

Furthermore, by relating aesthetic pleasure to perceptual intuition, Lee would avoid confusion between the pleasure or displeasure that may be found within perceptual intuition and that to be found through perception of fact. If one regards certain experiences as perception of fact, one might find the conditions of consciousness unpleasant, i.e. conditions to be avoided. Regarded aesthetically, however, simply as data of perceptual intuition, such conditions of consciousness might be sought or maintained, i.e. found to be "pleasant." As Aristotle has remarked, "for even when the original objects are repulsive, as the most objectionable of the lower animals, or dead bodies, we still delight to contemplate their forms as represented in a picture with the utmost fidelity."[1] Lee would account for such differences by dis-

[1] Aristotle, *On the Art of Poetry*, ed. by Lane Cooper (Ithaca: Cornell University Press, 1947), pp. 10–11 (1448 b 11–13).

tinguishing an aesthetic attitude from one in which moral, economic, intellectual, or other such considerations are of primary concern. Finally, conscious recognition of pleasure or displeasure as an evaluating factor occurs at the rather complex cognitive level of evaluation. Pleasure as evaluating factor provides the basis for comparison judgments of value.

Lee remarks that the words "taste" and "beautiful" both have reference to the comparison of values (*PAV*, 242). This point is related to his criticism of Kant and also to certain differences between his own aesthetics and the position which Santayana sets forth in *The Sense of Beauty*, a work which Lee follows on a number of points. Kant focuses upon taste and Santayana upon beauty. Of course the field of aesthetics can be restricted to such areas, but to do so would be to limit aesthetics to the philosophical study of comparisons of value. Following Lee's analysis, comparisons would be made by means of judgments of value, or evaluations, and the foundation of such judgments in the immediate apprehension of value would not be given a full account. Since evaluations are discursive, such a limitation might well facilitate the restriction of the field to an investigation of discursive forms, but consideration of the foundations of value within experience would be lost.

Furthermore, the comparison of value, or evaluation, might be more properly regarded as belonging within the domain of the philosophy of criticism. Stolnitz, for example, who makes considerable use of Lee's aesthetics in his text *Aesthetics and Philosophy of Art Criticism*, distinguishes between valuing and evaluation. He points out that the philosophy of art criticism is concerned with two activities, evaluation and criticism, and that the philosophy of criticism "deals with the process of *evaluation* – analysis and judgment of the aesthetic goodness of the object."[1] Lee provides a foundation for criticism in his analysis of aesthetic evaluation and valuation. His aesthetics offers more than a philosophy of criticism and rests upon an analysis of the epistemological foundations of discourse.

[1] Jerome Stolnitz, *Aesthetics and Philosophy of Art Criticism* (Boston: Houghton Mifflin Company, 1960), p. 372.

The field of aesthetics can be interpreted as dealing with the study of three very general interrelated problems: (1) the nature of aesthetic experience, which involves a theory of aesthetic value, (2) the philosophy of art, (3) the philosophy of criticism. In approaching aesthetics as the philosophical study of perceptual intuition and by developing the notion of evaluation as distinct from but based upon immediate valuational experience, Lee has dealt with the first and third of these problems.

Perception and Aesthetic Value includes a discussion of art. The word "art" is used to refer to a work of art, an object, rather than to an activity. Art or "aesthetic art" is described as "a vehicle which conveys aesthetic values" (*PAV*, 148). Although aesthetic value can be found both in nature and in objects constructed by man, those artificial products constructed especially for the purpose of being sources of aesthetic value are usually called "art" or "fine art." Art is "expression" in the sense that it is the expression of an aesthetic vision which the artist embodies in forms "which make it communicable to others (or to himself)" (*PAV*, 156). A number of issues regarding art are considered. Many questions in the philosophy of art, however, are left unanswered. It is not surprising, therefore, that Lee should return to the subject of art in his recent publications. In doing so he has modified and developed certain key points in his earlier aesthetics that are particularly relevant to the philosophy of art.

"Action, Perception and Art," which is based upon a paper read at a meeting of the Southwestern Philosophical Society in 1969, illustrates some of the changes which have occurred in Lee's basic philosophical position as well as a certain consistency in his thought.[1] In this article, Lee returns to the question of whether art is an activity or an object, a thing. He rejects his earlier position that "art" is to refer exclusively to an object and argues that "the two views of the nature of art (that is, whether it is an activity or an object) can be shown to be *essentially* related and not only externally so;

[1] Harold N. Lee, "Action, Perception and Art," *Tulane Studies in Philosophy*, XIX (1970), 55–63. Hereafter cited in text as "*APA*".

and that this can be done without blurring the distinction between them" (*APA*, 56). The basis for his argument is epistemological and leads to a development of his theory of perception. It hinges upon recognition of the inseparability of perception and action. "They can be distinguished but not separated" (*APA*, 61). Perception is regarded as a constructive activity; "hence the pleasures or satisfactions in perceiving are themselves pleasures or satisfactions in constructive activity" (*APA*, 56). This involves his explicit rejection of two assumptions made in the chapter on art in "*Perception and Aesthetic Value:* first, that nature is composed of separate or discrete things or events, and, secondly, that the data of perceptual intuition are ultimate in the analysis of the content of experience. These modifications can be attributed to Lee's increased concern with flux as a basic principle in epistemology, and also reflect the influence on his work of the philosophy of George Herbert Mead. The result is what might be described as a more sophisticated epistemological position with significant implications for the philosophy of art. The work of art is no longer to be naively classified as a discrete physical object and the role of interpretation in perceptual intuition is brought to the fore.

Although Lee does not approach these modifications directly in terms of the question of form, they can be interpreted as centering around the problem of form in aesthetic experience. They concern issues which Lee may have attempted to avoid in his earlier work. One of these is the metaphysical foundation of form. According to his later view, although he offers an epistemological analysis rather than ontological speculation, he does remark "there is no delineated or definite physical thing or object of any sort aside from the activity of an organism" (*APA*, 56). He holds, furthermore, that discreteness of parts is a condition of knowing and that perception is not the "passive mirroring of a ready-made world" (*APA*, 60). It would seem to follow that the formal organization of aesthetic objects and the individuation of such objects as separate from other experiential events would depend upon the activity of an organism. They cannot simply be regarded as the reflection of a structured external world of distinct physical entities. Of greater significance, however, is the role of form in connection with his later treatment of perceptual intuition.

Perceptual intuition is no longer regarded as the ultimate content of experience, but rather as being ultimate so far as *consciousness* goes. Lee now holds that data are intuited without consciousness, but, to be perceived, to become a part of conscious experience, an organic reaction or response to what is intuited is required. Lee calls this initial response which, though not conscious, is a necessary condition of consciousness, a "proto-generalization." He describes it as the proto-type of a concept. Such reactions occur with reference to stimuli regarded as *kinds* and require a "modicum" of reference to past and future, to stimuli that are not immediately present. They involve selection and interpretation. Perception is described as "the interpreting of the data of intuition in terms supplied by action and response" (*APA*, 60). Perceptual intuition is regarded as involving a "minimum" of interpretation. It appears, therefore, that Lee is introducing general forms into the perceptual process. A question then arises as to whether the introduction of such forms constitutes a loss of the distinction between perceptual intuition and perception of fact, between valuation and evaluation.

It can be argued that these distinctions are not lost. But, in order to retain Lee's distinctions along with his revisions of the theory of perceptual intuition, certain notions concerning the role of form in perception require clarification. In *Perception and Aesthetic Value,* Lee makes the point that perceptual intuition differs from mere sensation in that it involves form. He also remarks that objects are perceived in habitual ways and suggests that the actual conditions of perceptual grasp change as habit changes. In line with his later treatment of perception as dynamic and interpretative, we might more properly speak of perceptual intuition as involving *formulation* and thus avoid the more static term "form". The process of formulation requires reference to past and present, response to stimuli regarded as kinds. Forms as general "kinds", Lee's "proto-generalizations", are essential to this process. As Lee indicates in his article, the ability to respond set up in habits of response is one of the necessary selective factors in the interpretative process of perception. These habits include use of the conceptual and discursive forms that constitute the social inheritance of the percipient. This does not mean that the perceptual intuition *per se* is to be reduced to a given conceptual form, subsumed under a universal in this sense. It is not as intuition consciously discursive or judgmental. Considered analytically, various forms, including discursive forms, can be found within a given experience of perceptual intuition. Discovery of such forms may be part of the task of criticism as an intellectual explanation or rationale of evaluation, i.e. a value judgment. They are not to be identified with the aesthetic experience itself, however, which Lee relates to perceptual intuition and regards as immediate.

According to Lee, the original artist is "a person who both perceives and constructs more profoundly than others" (*APA*, 62). He can "interpret the stimuli now present (and of which he is not fully conscious until he has interpreted them) more richly and fully in terms of stimuli not present, that is, past and future" (*APA*, 62).[1] Some of Lee's remarks in this connection are reminiscent of Croce. For Lee also there are no "mute inglorious Miltons." But he apparently attempts to avoid the sort of "false idealism" of which Croce has been accused. He contends that the only criterion for knowing whether a person does perceive "more acutely and profoundly" is to be found in communication and the work of art is important in this connection. The artist constructs a vehicle of communication and his artistic vision and its means of communication are inseparable. These remarks suggest that the artist perceives in terms of his medium and in accord with the habits of construction that have become a part of his art. Lee does not commit himself as to the ontological status of this "vehicle of communication," the work of art. In line with his epistemological approach he restricts his analysis to the constructive activity of perception and indicates the criterion

[1] This point relates to Lee's more general epistemology which shows the influence of George Herbert Mead.

by which the artist is recognized. In so doing, however, he provides the basis for a communication theory of art similar to that suggested by his earlier work. His recognition of the role of interpretive, selective activity in perceptual intuition, however, permits the development of a theory of art as providing a much more complex mode of communication than that provided by his earlier analysis.

Lee's emphasis upon the intrinsic value of perceptual intuition in *Perception and Aesthetics Value* suggests a sort of pure aesthetic which would lead to a philosophy of art as isolated and divorced from the complexities of life. Such an art would have little to communicate beyond the perceptual pleasures found within it. Although such purism is modified by his theory of "derived values" which, following Santayana's theory of expression, would allow for nonaesthetic values to be found within aesthetic experience, the incorporation of such values is not fully accounted for. Lee's more recent theory would allow for a variety of values implicit within the selective, interpretive activity of perceptual intuition. By providing such an intuition the artist can make one aware of these values. He can communicate at a non-discursive perceptual level by means of aesthetic valuation. Such communication is non-cognitive in that it is intuitive and does not consist of literal propositional discourse. It might be regarded as cognitive, however, if perceptual intuition is a way of knowing, although Lee does not use the term "cognition" in this sense. The artist cannot be dismissed as merely providing a source of perceptual pleasures. Insofar as the original artist "perceives and constructs more profoundly than others", which includes interpretation of the social structures implicit within perception, he is a major source of awareness. He is essential to the well-being of society, especially the complex, rapidly changing society of our own era.

Furthermore, it seems to follow from several of Lee's remarks that the artist is not to be classified as a distinct kind of human being essentially different from his fellow man. The difference is a matter of degree rather than kind. If perception itself is regarded as a constructive activity, this creative action is characteristic of man as a perceiving animal and aesthetic value as found within perceptual intuition is basic in man's relation to his environment. Aesthetic experience is as fundamental to man as perception itself and aesthetic value as found within perceptual intuition is basic in man's relation to the world in which he lives. The aesthetics of Harold N. Lee not only offers useful distinctions in the analysis of aesthetic experience and aesthetic value, it includes a philosophy of art and provides a foundation for criticism. What is more, it contains the keystone for a more general theory of philosophic humanism.

ON THE PHENOMENON OF OBLIGATION

EDWARD G. BALLARD
Tulane University

Why is one obligated? Or, indeed, is one obligated?

These questions were discussed once in prison. Crito awakened Socrates at dawn, and they quickly fell to debating whether Socrates ought, against the legal decision of the City, to flee Athens, a possible and – as Crito saw it – attractive alternative to drinking hemlock (*Crito* 48B; 50A). Socrates concludes that he is obligated to the Athenians by a contract to carry out their wishes, a contract implicit in his relation to the city which had fathered him (48B). This contract then seems at the end of the dialogue to become the Ideas, those intelligible realities in whose image both he and the city desire and ought to desire to shape their characters. I cannot, of course, even begin to present and discuss examples of the whole range of the phenomena of obligation in the space available here. Nevertheless, this Platonic example is typical certainly of an extremely important class of obligation experiences, if not of all. We shall give our attention exclusively to this type and attempt to disclose the logos of which it is the appearance. This limitation will require, I think, no further mention.

The Socratic illustration suggests a convenient formal mode of expressing this kind of obligation phenomena. Generally we may say that A is obligated by B to perform C (although the performance of some non-C is possible and perhaps more attractive). Obligation, then, is the relation holding among A, B, and C. When we say that Socrates is obligated or feels obligated, we are saying that he is the A in this relational complex. In my attempts to characterize this relation more concretely, I shall find it necessary to refer frequently to the negative of obligation, for instance to the experience wherein the demands of obligation might have have been met but in fact were not. This is the experience of guilt, an experience intimately related to awareness of self. As Socrates repeatedly points out he would not have continued to be himself had he shirked his duty and fled Athens.

Normally for one belonging, as Socrates claimed to belong, to a rational tradition, reasons are expected to be given for believing that the relation of obligation holds in the manner in which it is asserted to hold. However, these reasons do not seem to require a separate entry in our formulary, because the justification of such beliefs is already intimately linked with the B by which one is obligated. It would appear reasonable to expect to find the reasons for being obligated precisely within that by which one is obligated.

The problem of obligation, then, is the problem of understanding, of giving appropriate meaning to, the elements and relations of our formula. We shall, therefore, have to be concerned with (1) the question of the definition of the relation denoted by 'obligation'; (2) the question of the identity of C, the obligatory actions themselves;[1] (3) the problem of the identity of B, that from which obligations arise and by which they are justified. For the limited purposes of this article, emphasis will be placed upon the third of these tasks, the question of the source and justification of obligations. This is the problem of determining the nature and strange power of the kind of obligation with which we are here concerned; this power is such, it should always be borne in mind, as to lead one upon occasion to sacrifice one's own life.

I. *The Older Self and the Feeling of Guilt*

It will be sufficient in order to make a beginning, to indicate the genus of obligations without providing this beginning with an elaborate defense. Obligations, we say, are valuations of a certain kind. Valuations I shall think of as visions of value, predictions of satisfactions,[2] which are to be verified

[1] These obligatory acts or valuations are often termed simply obligations. 'Obligation' also names the relation among A, B, and C. This ambiguity need cause no difficulty once it is recognized.

[2] Value will be considered to be the satisfaction of need. If these terms be allowed sufficient generality, no violence will be done to the variety of human interests, causes, and devotions. Harold N. Lee offers a more general definition

or falsified in the experience of the actual event. Thus valuations are motives to acts. As it might be formally expressed, A performs C because of certain valuations. Now to be obligated is at the least to choose and adopt a certain class of valuations; it is to commit oneself to the pursuit of values of a certain kind. This commitment (to B in our formulary) is not a matter of the comparative indifference nor of mere preference with which many value choices are made. Rather, on its positive side, this class of valuations is evaluated, usually quite spontaneously, at least on the part of persons well attuned to the given culture, with unique force. In addition, this evaluation possesses a rather peculiar negative power. Violations of obligations are thought to place the violator in dire, if frequently vague, jeopardy. This peculiar force of obligatory valuations – both on its positive and negative sides – is the difficult thing to understand. How, indeed, is the persuasive power to be identified which differentiates obligations from other valuations? Why, that is, are we obligated?

Let us say that A holds himself obligated to B to perform C if the consequences of the non-performance of C is a sense of guilt in A.[1] Now this expression would appear to provide an empirical test by which obligations (acts or valuations substitutable for C) might be detected. And in many areas and eras, especially religious eras, it has been accepted in just this manner. "Let your conscience be your guide," it is confidently said, for some recollection has remained of the etymological sense of 'conscience'. Conscience in this sense refers to knowledge (*scientia*) which is had in common with someone else. In the traditionally religious instance this someone else is said to be God, specifically the father-God, who reveals the standards of human behavior and who punishes and rewards accordingly. This God is the B in the formula above, an

of value: value is "a quality imputed to an object by virtue of the fact that it can come into a certain relation with a sensibility" (from "A Precise Meaning for Objective and Subjective in Value Theory", *Journal of Philosophy*, XXXVII [1940], 626–37). If his "quality" can be stretched to include my "satisfaction," then his definition can include mine as a species.

[1] I assume that anyone who is curious about obligation will be able to provide an ostensive definition of guilt, that sense of blameable deficiency in which the person himself is peculiarly involved. In the present section I do not consider the positive correlative to guilt, the sense of moral self-approval.

authority who ties obligations onto men by the thongs of a sense of obligation and guilt which he mysteriously but systematically elicits in his creatures.

A little travel in foreign lands where men having a differently trained conscience live under other gods who prescribe different standards has often been sufficient seriously to undermine the conscience of a people. Likewise it is disturbing to the confidence sometimes placed in this bond between god and man to observe the sense of intolerable guilt which leads the psychopathic personality to confess to a murder which he never committed or which leads the religious recluse to believe he had crucified the Christ within him because he looked a little longingly after a woman's shoe. If the pangs of conscience can and do vary in this fashion both with the social and individual milieu, then they may indeed be a misleading clue to the nature of obligation and the authority which imposes it. But this familiar conclusion is too quickly drawn. There may be an element in common among many of the occasions of sensing guilt which, if it can be disengaged and expressed, will prove to be helpful to understand.

A step toward disengaging this common element can be made by comparing the sense of guilt with certain other feelings to which it is closely akin. It is similar to and is sometimes confused with embarrassment, for example, or with a sense of shame, wounded pride, or the confusion which follows upon the commission of a mistake. It will be profitable to contrast guilt with the last mentioned emotion, the feeling which accompanies the recognition that one has made a mistake. We approach this matter by recalling a psychological theory concerning the development of conscience which is held to account for the sense of authority with which obligations are imposed; we develop it in order to account for the diversification of the sense of guilt from other feelings of the social animal.

Freud, and others as well, have concluded that conscience is acquired in consequence of the operation of the natural tendency, which he posits, to imitate or to "identify" with certain others. The child imitates other people and thus absorbs them or aspects of their behavior, into his own personality. Among those whom he imitates in this manner are other persons who are engaged in observing and criticizing him. He imitates this criticism of himself. Thus, eventually he comes to acquire an observing and critical attitude toward himself and, like his father or other authority, to punish himself for that which he has adversely criticized. Primarily, he introjects or incorporates his father and is said in consequence, to acquire a set of internal standards or a conscience which actively opposes and controls his primitive impulses. The admission that some act or omission is to be adversely criticized is registered emotively as a sense of guilt. To the child, who makes no fine distinctions and tends to react wholly or not at all, this sense of guilt is not related so much to a specific act or error as to the self as a whole. The child does not say to himself, "This action was wrong and should be corrected." Rather he says, "I am evil and deserve punishment." The child makes no distinction between awareness of a mistake and sense of guilt.[1]

[1] Freud has speculated in an interesting history-myth upon the initial act in the formation of conscience. He thinks of it as an actual act of (often repeated) cannibalism in which a group of the more ambitious brothers in the primitive horde banded together, driven to extremes by continual frustration of their sexual and other interests, by the powerful and god-like father (*Totem and Taboo*, trans. by A. A. Brill [New York: New Republic, Inc., 1927]). They murdered and ate the father. This act was not the simple removal of a barrier to gratification. For the father was not only the powerful protector but, after a fashion, he was also loved and admired. Eating him was the way of acquiring his mana, of becoming the father, but only at the price of annihilating the loved figure and the tabooed embodiment of power and security. Thus, the sons incorporated the father and acquired power to satisfy appetites, but only at the expense of suffering guilt. Thus incorporating the father and taking responsibility for themselves awakened ambivalent feelings.

This theme has been elaborately developed by Theodor Reik in (what I take to be) a prose epic (*Myth and Guilt* [New York: Braziller, 1957]) which elaborates the Christian doctrine of the identity of the Son with the Father by finding an identity at a primitive level of mind between the father of the primal horde, the tree of knowledge of which Adam ate, and the cross of the New Testament which Christ, as it were, absorbed into himself. Then in addition, it finds an analogous identity between Adam and Christ. The whole is the story of the son's attempt to become the father by actual or symbolic eating and thus to become the father-god, and of his recognition of the nature of this impulse and his consequent guilt, sacrifice, and atonement. To speak of this attempt as the source of an original sin of mankind is to claim an inevitable and universal development of a sense of guilt in the route to maturity which men take. There is good evidence that such a sense does tend to develop, whether Freud's myth and Reik's development of it point toward its inception or not. I allude to it later in terms of an ontological defect.

Freud looks upon this development as universal in tendency, if not in fact, and finds it to be the source of a paradox of our culture. For if one attempts to satisfy his often murderous impulses directly and forceably, then he is punished by society, and if he voluntarily frustrates them in the interest of gaining the love and approval of others, then he is still aware of the wish to gratify them at the expense of others, and hence he feels guilty and punishes himself by further self-inflicted frustrations in order to rid himself of these feelings. In either event he is punished. Freud thinks that civilized life will become more and more complex, and the occasions when one must choose between criminality or self-frustration and self-punishment will become more and more numerous until at some point the tension between impulse and conscience will become too great for men to bear and they will revert back to type.

This dire prediction may not be the only alternative, for note that the incorporation of the father as the interior power and ruler has already set up in the child an internal model of the family of which he is a part, thus providing a first instance of the Platonic analogy between the individual and the group. In the movement to adulthood further diversification of the psyche occurs in consequence of the introjection of other and differently functioning individuals. One might build a case for supposing that not only the ruling father but also the hunting and fighting older brothers as well as the agricultural and weaving women acquire their representatives within the psyche of each normally developing child. As society passes through the stages of ruler-selection, division of labor and specialization for protection and production, so the individual who initially tends to incorporate the social authority will also find himself incorporating other functioning parts of the social structure. Thus he will come to mirror in his personality the functional divisions of society. As the society grows in complexity and clarity of its organization, so we must suppose, does the individual. At the least, he develops not only an internal ruler, but also an internal artisan who uses his skill to produce satisfactions for his needs – more or less in harmony with the internal rule – and likewise an internal protective and executive function which mediates between ruler and artisan and protects the whole from internal as well as external injustices.

The upshot of this development is that the person need no longer feel his whole self to be guilty for any chance error, as these are measured by the complex and graded civilized standards, but on some occasions he need feel only that something went wrong with some subordinate aspect of his functioning. In effect, his sense of a pervasive guilt consequent upon admission of error is dispersed over many functional parts of the self, just as the tasks of providing for society's needs are distributed among many specialists. Error thus becomes less personal and easier to bear; compensation for it is more varied. One important product of civilization, perhaps a saving element, may be just this diversification of error in relation to the more complex self and differentiation in the emotional reaction to it. Not all errors, only some, are violations of obligation. Which are these?

Some errors have always been connected not merely with some aspect or specialized function of the self but rather with the very identity of the self. Let us express this important distinction as the difference between sense of mistake and sense of guilt. One feels a sense of guilt when the acknowledged error violates or conflicts with one's identity. That is, one feels guilt when the acknowledged error is associated with the self's violation of his internalized ruler, the father within him, rather than when the error is attributed merely to some subordinate aspect of the self. But if the actor's identity is not called into question by the error, then he is aware merely of a mistake. Our obligations, then, are valuations or acts which are related to guilt understood in the sense indicated.

It is, of course, very difficult to make a clear distinction between the self and aspects of the self; here no claim is being made fully to define the self. It is sufficient to note that the self's relation to the parent and parental authority is certainly a basic relation, one without which the person would not be. Naturally this relation figures fundamentally in the psychic economy. So intimately, I suggest, that one's sense of self is even identified with the internalized father-authority. This is the ancient self, but a self which is also contemporary. It may be, as we noted, that any occasion of sensing guilt is connected with the atavistic impulse to incorporate the father by actual or symbolic eating or the impulse to absorb him, his power and his function, within one, and through this violation to become the father-god; any impulse or act which reawakens the feelings connected with this original impulse is felt as guilt. Thus the sense of guilt is felt when the very self is taken to be the culprit. This self, I suggest, is the B element in our formula and the element common to guilt feelings which lends to them their peculiar force. And this it is which lends power to obligations. When error is attributed not to the self but to some aspect of the self, to the self as soldier or as artisan for example, then one is aware not of guilt but of a sense of mistake, the uncomfortable feeling of an error in technique.

II. *The Future Self and Actual Guilt*

Up to this point I have been concerned chiefly with the feelings or emotions experienced upon occasions when the sense of obligation is violated. The ensuing sense of guilt was related to one's particular self-evaluation, which in turn was held to be derived from ancient and deep-seated emotions originating in parental evaluation. But the sense of guilt is a feeling, and feelings are particular and notoriously variable. It will be desirable, therefore, to free the concept of guilt from its connection with feeling and from the mytho-psychological scaffolding which was needed initially to relate the sense of guilt to the self. Now we need to express its universal character. For clearly, to know that people are apt to feel guilty on certain occasions does not provide what we seek, namely an answer to the question why we are obligated to avoid guilt by performing certain acts or adopting certain valuations. All a psychological account can do is to tell us that by the operation of an effective mechanism a set of possible choices is in fact placed before us in a powerfully persuasive manner. But an account of this sort does point to relevant data. We may now move a step further toward a general account of obligation by the aid of the notions of ideal person and hypothetical self.

The notion of ideal person is implicit in conscience. One feels oneself responsible to the father or the authority component in conscience to become precisely the person which is not yet: the ideal person. This ideal person may not be clearly described, but, buried under tradition, convention, and formularies, it is nevertheless an essential part of every person's cultural inheritance. Occasionally this ideal person is hatched from its cocoon of convention and is expressed with some care and even criticized. When the ideal person implicit in a family, tribal, or cultural tradition is rendered explicit, I shall call it the 'hypothetical self'. We may now define actual guilt as distinct from the sense of guilt. When A holds himself obligated to B to perform C, and the failure to perform C, when C was possible, is followed by the recognition of a specific disparity between the actual self (A) and the hypothetical self (B) – which would not have existed had C been performed, – then A has recognized actual guilt. The sense of guilt, then, is an emotion which accompanies the (intellectual) awareness or recognition of this disparity. This definition takes the emphasis off feeling and emotion, a move which is desirable since guilt feelings may be absent when actual guilt exists and vice versa.

The B in our formula has now been identified as the hypothetical self, the clarified and criticized cultural ideal, the ideal father which the son as a man of his culture, ought to aspire to be. The standard by which those valuations which are to be called obligations are selected and measured is the hypothetical self. Thus, obligations are self imposed. Whatever the psychological mechanism by which we are persuaded to accept obligations, it seems to be obvious that no external authority or force can impose them. Obligations, if they figure in morals, are a class of choices, and only the actual and responsible person can make his own choices.

The transition from the first section which we have made is the following: as the sense of guilt is to the feelings most intimately associated with the self so actual guilt is to the hypothetical self. This transition provides us with a definitional measure which, relative to the culture whose standards are being used, distinguishes between genuine or actual guilt and pathological guilt. It enables one realistically to criticize oneself by comparison with an explicit ideal. The point emerges again that guilt, whether guilt feelings or actual guilt, is intimately associated with the self and is aroused when the identity of the self is brought into question, rather than when some minor activity or engagement is involved which is not thought to touch upon the identity and unity of the self.

The definition which has just been offered is quite general, for it can be shown to be meaningful in any cultural or doctrinal context. This generality obtains because the notion of hypothetical self (the B in our formula) is taken as a variable which may be given content by indefinitely many doctrines. The other variables are evidently dependent upon this one. The question inevitably arises concerning the limits on the range of the variable B. I suspect that this question is inevitable, for it bears on what seems to be a persistent longing to transcend cultural and doctrinal limitations. If obligations are self-imposed, what, men have always desired to know, is this self which imposes and accepts obligations? A considerable part of our tradition is at one in the conviction that any formulation of the hypothetical self, whether implicit or explicit, is inevitably distorted in consequence of a native human inability to see or to understand one's self in a final or complete sense. One recalls Socrates' injunction to know oneself together with his protestation of ignorance. Then there is the Christian recommendation to imitate Christ, who at the same time is said to be a mystery beyond man's understanding. Finally should be recalled the conviction that man is primarily the drive to transcend any concept of his essence, any identity which he may adopt. Obligations to the ideal self, then, are incorrigibly infected with a kind of uncertainty in virtue of which any act, however virtuous seeming, may turn out to be inappropriate, a thing of guilt.

This important point may be made in other ways. I have already indicated the ambivalent attitude in which obligations are held by remarking on the identity of conscience with the father; but the father is a powerful and a dangerous being to challenge and internalize. This being, however, is internalized as the ideal part of the self, the ineffable self which one both fears and strives to become. The thesis of the incomplete intelligibility of the individual gives expression to a not dissimilar suspicion that any given theory of the individual will distort him, perhaps tragically, in the direction of the aspect within which the theory is developed;[1] and yet, paradoxically, intelligent action can scarcely go forward without some vision of the self that was and is to be. Not the least important presentation of this radical but unavoidable self-uncertainty is offered by the Greek doctrine of tragic hybris. This doctrine, moreover, hints at a union between the hypothetical self which one feels obligated to become and the more ancient self located in the conscience which obligates, and thus it suggests a more complete, a more real, self.

III. *The Real Self and Ontological Guilt*

Although the hero of a Greek tragedy was a superior person, it will be remembered that he always made a decision about himself or the office his duty required him to perform which was out of character or unrealistic in that it demanded of him greater knowledge and skill than he actually could command. Only after being tried by fate does he come to the realization that he has misidentified himself. Hybris is the tendency of every person to over-estimate himself and to attempt the deed which he can not perform or to assume the identity which is not his own. And indeed, not the Greek only but Western religions and myths quite universally attest to a natural blindness, an essential lack, in the human being by reason of which he cannot but fail in reaching his highest satisfactions. One's highest obligation may be to become oneself, but efforts directed toward self-realization are so consistently frustrated that one is led to suspect the presence of some ontological impediment in the path to self-identity. As the myths of man's fall insist, it is as if one were already guilty, being human.

The psychological myth alluded to in the preceding section conveys the same point when it speaks of the rebellious sons

[1] Cf. my article, "Individual and Person," *Journal of Philosophy and Phenomonological Research*, XVIII (1957), 59–67.

who try to absorb and become their father. In fact, though, they are not their own father. But the over-weening tendency to regard oneself as the center of the universe persists. Consider, for an instance, the philosopher who flatters himself that he has incorporated and expressed reality itself in his systematic formulations. Or consider the moving spirit of our own times, so confident of the progress achieved by one generation absorbing and building upon another, moving to the superman-technician at its summit; it suggests nothing so clearly as the irrationality of the tower of Babel.

To see this insolent and unrealistic element in the human nature itself, aspiring to god-like domination, would seem to be the disclosure of despair. Still as the hero moves from catastrophe to catastrophe a direction of movement becomes discernible. After each insight the hybritic defects of each hypothetical self are recognized and to some degree apparently removed as the protagonist reformulates his beliefs about himself. Every tragedy is a self-criticism and a partial self-clarification. And yet complete clarity is never achieved, never achievable. Wisdom, as Plato observed, is for the gods only.

In a similar vein, we may add that the willingness of philosophers to criticize each other by reference to some impalpable ideal of the complete wisdom, the perfect man, the ideal language, the unmuddled concept, is an admission of their hybris and an attempt partially to correct it. Among the Greek plays, this movement to clarity is most evident in *Oedipus at Colonneus*, and the *Orestia*. It appears to be suggested that there exists a real self to which the expressed hypothetical self approaches somewhat closer after each tragic recognition of past deficiency and ineptitude. This real self, this tower beyond tragedy, if only one knew where and what it were, might be used to take the measure of the hypothetical self and thus to define once and for all what obligations there are. But this self seems only vaguely to be dreamed in the inspirations of the prophet and related in stories about the Hero, the Superman, Nirvana, Sainthood and friendship with God. The Romantic philosophers despaired of it while pusuing it. The Positivist easily puts it down as a ghost story.

If our obligations could be justified only in terms of inspiration and poetry about the real self and its ontological defect, they could be challenged and devaluated by a sophist's criticism. It is easily observed that poetry about the real self does not offer a convincing answer to the problem why one is obligated or why one is guilty. Also anyone might admit that all men have always held themselves obligated to do their duty and admit, further, that he is a man; still he could quite logically ask; "Why should I choose to obligate myself thus?" To point out that by doing his duty he avoids or minimizes guilt and becomes his self-approved or even his real self will merely force him to put the question again.

An easy issue out of this impasse is to assume arbitrarily some one obligation, some one imperative which will be directive of choices. If the assumption is aptly selected, one's ethic may still be saved. And this procedure is susceptible of a logical defense. First principles appear to be of the nature of assumptions; and if there are to be imperative conclusions, at least one imperative principle must be assumed. This procedure, however, is too luxuriantly fertile in ethics, for it can be used to justify any set of so-called obligations whatsoever. Accordingly many philosophers have been dissatisfied with it and have sought to show in some fashion that the primitive obligations which they assume ought to be assumed. Usually they argue that among the many imaginable hypothetical selves one of them ought to be regarded as the real self because of certain rather persuasive qualifications which it possesses. Among these qualifications would be the minimizing, so far as possible, of the ontological defect which tradition ascribes to man's nature, the defect which entails that any decision concerning his identity will be faulty.

In particular, the classic tradition, in defining man as the rational animal, has sought cleverly to define its hypothetical self in such a manner that it can use this definition not merely as an ethical primitive but as a justified or non-arbitrary ethical primitive, one, moreover, which can be counted upon to lead most directly to self-knowledge. Perhaps the rationality of a person is only one of his aspects and perhaps, in identifying himself with it and developing primarily the potentialities of this one aspect (as if the individual were or were to become completely intelligible) is a tragic decision. And yet one must admit the peculiar and forceful persuasiveness of this identification. Without development in just this regard, tragedy would not be possible, for awareness of the tragic hybris, or recognition of the gap between the actual self and the hypothetical self as revealed in the trial by fate, is precisely a rational recognition. Without this use of rationality no distinction between the actual, the hypothetical, and the real self could be recognized or expressed. To put the matter somewhat paradoxically, it is only by identifying himself primarily as rational that a man can find out that he has misidentified himself, if he has done so. My point is that the identification of oneself as rational is unavoidable, even if the identification be faulty.

It is easy to see that the traditional convictions about the primary obligation of each man, from which his particular obligations are derived, refer back to this identification of him as rational. The primary obligation of each person is to pursue the being of himself through the rational process of hypothesis, trial, and reformulation of the hypothesis. Inevitably the recognition of a dis-harmony between what one actually is and what one naturally and more happily might have been can not be accepted with self-satisfaction nor logically acquiesced in if one also accepts the conviction that one's nature is rational, and to be rational is to be one's coherent and unified, real self. Otherwise expressed, it would be self-contradictory for a rational being to accept as final a disparity between his actual self, his hypothetical, and his real self. Thus he accepts the rational self as his real self. If on the contrary one rejects the definition of oneself as rational – and seriously undertakes to put this decision into practice – one will be in no position to judge the outcome of this practice. Such a person will never be in a position to know whether his choice of means achieved the end which he posited. In consequence of some such reflection as this the individual is persuaded to risk a definition of himself from which those obligation follow that dignify him as a rational person.

Thus obligations are recognized and exist for us because

they follow from our acknowledgment of our rational nature. It may sometimes appear difficult to say whether the conviction of rationality is derivative from awareness of guilt and obligation, or whether the logical order runs in the other direction. Or it may be that these are equivalent ways in which a fundamental intuition of the self is expressed. However, the important point is susceptible of formulation: the actual person (A) is obligated to his real and rational self (B) to perform C whenever the omission of C would prevent his recognizing the disparity between the actual and the hypothetical self or between the hypothetical and the real self. This contemplative ideal, or something like it, has some claim to being the trans-cultural faith on which much of Western history has hung.

Still it must be remembered that C will to some extent inevitably be omitted, such is the consequence of the limited rationality of the creature owning to the ontological defect. Failing, thus, by his very nature to see or to be himself or to see just how he is not himself, this defective creature may be said to be ontologically guilty. This is the ironic aspect of the faith to which the West has clung.

IV. *The Ambiguity of the Real Self*

This conclusion and the path by which it was reached prompt a few observations upon the nature of the real self which are relevant to our theme. Obligations were specified as a set of valuations to which the person is committed both affectively and in principle; these were considered to be the self or an essential component of the self (the B of our formula). Yet clearly the standards which judge the self are not identical with it, although that which accepts and imposes these standards may be the self. Neither for that matter are his impulses, which are judged, identical with the self. Then an argument was adduced which concluded that the classic definition of man is a very persuasive definition; the obligations of men are to be recognized as rational obligations. They are accepted by the rational self and imposed because they are rational. Only in thus accepting them can the self become and remain itself. If, however, history has made any one thing evident, this is that rationality is an ambiguous notion. Men of almost all times and places, allegiances, and varieties of commitments are insulted when they are called irrational. Is the rationality which they severally desire equally various? Is it possible, we are led to wonder, to define rationality in a positive and unambiguous manner?

Here we cannot pass the whole history of reason in review, but it is important to observe that even to a cursory inspection of Western history two broadly differing conceptions of the nature of rationality have emerged. The older and more complex notion is derivative from the ancient Greek conviction that man's rational nature is in some sense dual, being characterized by poetic powers of intuition and insight as well as by ratiocinative capacities. The difficulties in coming to an understanding of the former psychic powers are certainly enormous, and yet a long and persistent tradition has held that these powers exist and are, along with one's ratiocinative powers, one of the indispensable means for working out one's fate. Indeed, insight into the perfect life possible for humanity was claimed for the poetic dimension of reason and expressed by Plato in poetry about the chariot of the soul (in the *Phaedrus*). At the same time, reason also admits to its inchoate tendency to hybritic error (its ontological defect) concerning its particular identity. The simpler, and in the contemporary world the more prominent, conviction about the nature of mind eliminates the ancient poetic psyche and tends to regard the discursive and ratiocinative intellect and its aids as exhaustive not only of the mind but of the self. The metaphysics of this view, if elaborated, would probably side with Leibniz in holding the individual to be intelligible through and through. This belief about the nature of the rational individual lies close to the surface of the Neo-Classic admiration of clarity and common sense as well as of the modern valuation of the scientist. The scientist-technician, undoubtedly, is the modern hero whose dealings with fate are regarded as exemplary; the economic and political worlds form his thiasos. This modern self admits to no hybritic defects and expects, through the continued application of reason and the scientific method, to enjoy ever increasing prosperity.

We may look upon these two conjectures about the nature of the rational self as two possible decisions. The first and more complex one appears to lead along a negative route toward identifying the person primarily through cultivating recognitions of his hybritic self-misidentification. The second would ignore this possibility and would turn rational activity toward problems related to the world or toward what Kant calls the empirical self with the expectation of discovering positive knowledge about, and finally power over, this subject matter. The former is the tragic way, the latter the way of Progress. In spite of social pressures, it may still be possible for individual persons to weigh these alternatives in the light of their insight into valuations and obligations and to arrive at a rational decision, – although such a suggestion is not innocent of irony.

It is undeniable that the second of these two views of the rational self has been productive of much which we would not willingly surrender. At the same time its period of dominance has been curiously coincident with the problems and confusions, the wars, tyrannies, complacency, and frustration with which the last two and a half centuries have so intimately acquainted us. Is there a connection? There may be. I cite as evidence only one thing: the general moral concern and especially the expressions of guilt sensed by specialists as well as by ordinary people upon the occasion of the Hiroshima bomb explosion. The inescapable suggestion was that some essential matter must for too long a time have been left out of the calculations. I suggest also that an inspection of history indicates that the tradition which has guided us longest through the mazes of time has been the very tradition which has clung to the dual views of the nature of rationality despite its difficulties and sometimes its negative consequences. This is the tradition which has been the more specifically concerned with human beings and the more successfully conservative of human values. It is especially to be noted that the problem of reconciling humane studies and human values with scientific ideals and pursuits persists only in the context of the more characteristically modern of the views just mentioned. In the classic tradition there are problems enough, but one matter it has settled: the relation of an impersonal knowledge of nature and the economic wealth which it can produce to specifically human values and obligations is and clearly ought to be the relation of means to end. Without this subordination no civilization, no self – at least no classic self – is possible.

Shall we, then, reject the dangerous invitation to Progress and settle for defining obligation within the safe confines of traditional reason and virtue? Let us consider again the path we have come. Within the Socratic type of the experience of obligation we have seen that there is an inner or psychic compulsion to accept obligation because failure to do so exposes one to guilt. Guilt at that level is the emotional reaction to possible loss of identity, identity here being understood in terms of the father-son relationship. Further and at another level, we saw that obligation tends to be accepted since failure to do so threatens one's hypothetical self or cultural and social identity, or else, finally, this failure may threaten the attempt to be or become one's true or real self. But this real self is said by tradition to be the rational self. Rationality, however, was observed to be ambiguous. Is there, now, some authoritative rationality to which appeal can be made in order to select the appropriate sense of rationality? Rather than engage in some such dogmatic appeal, let us make a final generalization from the three grounds of obligation which have just been reviewed.

At each level of response, failure to acceed to obligation was met with some threat to self-identity. In general, it may be said that the self is obligated to the self to be the (real) self, or else to suffer that one's self-identity be brought into question. But since the real self, which is the final ground for accepting or rejecting any obligation, is the rational self, and rationality is an ambiguous notion, at best an historical and changing one, the identity of the self will always be in question. Correlatively, if guilt be thought to be the recognition that the identity one has adopted is questionable, then one is always guilty. To accept obligation as a problem, we must finally conclude, is to recognize that the identity of the self is always at issue. Or perhaps one may say that the basic property of the self regarded as rational is self-questioning. In other words, the logos of the phenomenon of obligation is self-questioning. Less acceptable alternatives to this conclusion are identification of the self with a specialist rationality which leads only along the destructive paths of Progress or a more classic rationality which holds one within the tried but too well explored limits of traditional virtue.

POSITIVISTIC PATHS TO VALUE

ROBERT C. WHITTEMORE
Tulane University

It is stylish today to say that so-called moral judgments are not judgments in the true sense. Some say that they are volitional utterances. Some say that they are emotive utterances. The present essay will attempt to show that moral judgments are judgments in the correct sense of the term. It will also attempt to be precise about the volitional and emotive references of moral judgments.[1]

We begin with a question: What sort of an ethics, if any, is possible granted the validity of the Wittgensteinian principle of verifiability and the threefold classification of sentences as significant, tautologous, or meaningless?[2] Three, and only three, alternative types of ethical theory seem possible given the above conditions.[3] These types I label (i) Emotivist, (ii) Positivistic Utilitarian, and (iii) Significant Tautologies.

The first type is the outgrowth of the conclusion that ethical statements must be expressions of emotive states since they do not appear to come under the classification of sentences as significant, tautologous, or meaningless. The second type stems from the fact that the utilitarian criterion of value implies the applicability of the principle of veri-

[1] Harold Newton Lee, "What Are So-called Moral Judgments About?," *Tulane Studies in Philosophy*, VI (1957), 87. Hereafter cited in the text (in parentheses) as "*WMJ*," followed by the page number.

[2] Significant sentences say something about the world of fact. Tautologies say nothing about this world but may have meaning. Meaningless or nonsensical sentences are not really sentences at all but mere juxtapositions of words which, if they do not violate the rules of syntax, give the appearance of sentences.

[3] Actually there is a fourth alternative, that is, to deny that any type of ethical theory is at all possible given these conditions. When Wittgenstein declares, "there can be no ethical propositions" (*Tractatus*, 6.42), he seems to be expressing such a view. Since, however, such expression must be regarded more as an attitude toward ethical theory than as an alternative type of ethical theory, our three alternatives may be said to exhaust the relevant possibilities.

fiability to ethical judgments, while the third type is the result of the recognition that ethical judgments may have significance as tautologies even though they lack significance as sentences.[1]

Under type one (Emotivist) I subsume all theories whose basic thesis is the reduction, partial or complete, of all ethical statements to statements expressive of the emotional state or psychological reactions of the subject voicing the statement. I propose to examine three such theories (those of A. J. Ayer, R. Carnap, and C. L. Stevenson) in more or less detail.

Under the second type (Positivistic Utilitarianism) I include all theories whose criterion for value of any thing or act is the material utility of that act or thing. I shall examine one such theory, that of Moritz Schlick, and adumbrate another of my own.

I am unfamiliar with any writing which would serve to illustrate type three (Significant Tautologies). Conceding the possibility that tautologous statements may possess significance, it remains, nonetheless, my conviction that no ethical theory which takes its start in the admission that all its statements are purely analytic in character can be considered really satisfactory. All significant ethical statements must be, I submit, synthetic in character whether that significance lies in the expression of emotive states peculiar to the individual voicing the statement or in the assertion of the material utility of the thing or act in question. I shall, therefore, restrict my discussion to types one and two. A brief recapitulation of my reasons for preferring type two to type one will conclude the paper.

[1] "Significant" is used here, and throughout, in the special Wittgensteinian sense of the term. That is, significant sentences are, by definition, sentences which admit of the possibility of empirical verification.

II

> When philosophy deals with subject matter which is close to our ordinary experience (and surely, value theory falls here), its theoretic conclusions must be clearly and unequivocally applicable to experience; else it is not performing its task.[1]

Emotivism presupposes that we are not voicing a "significant" sentence when, for example, we assert that: X is right (or wrong). In its most extreme formulation (that of A. J. Ayer) the emotive theory maintains:

> In saying that a certain type of action is right or wrong, I am not making any factual statement, not even a statement about my own state of mind. I am merely expressing certain moral sentiments ... In every case in which one would commonly be said to be making an ethical judgment, the function of the relevant ethical word is purely "emotive." It is used to express feeling about certain objects, but not to make any assertion about them.[2]

This extreme form of the emotive theory is not widely held today. It is not difficult to see why.

Ayer's statement maintains that when I say that X is right (or wrong) I am not making any factual statement, not even a statement about my own frame of mind. In other words, I am not even indicating my personal approval or disapproval of X. My assertion that X is wrong must be understood to be nothing more than an unmotived emotional ejaculation, involuntary and unintentional. Yet this seems clearly inadequate. Ethical statements, whatever their status, are patently more than involuntary emotional ejaculations. When I declare that X is right (or wrong) I seem at least to be expressing an emotion of approval or disapproval with regard to X. Yet on Ayer's view I cannot even be doing that, for if I were, such an expression would seem to imply some attempt on my part either to communicate my feeling about X or to

[1] Harold Newton Lee, "The Meaning of 'Intrinsic Value'," in R. Lepley, *The Language of Value* (New York: Columbia University Press, 1957), p. 178. Hereafter cited in the text (in parentheses) as "*MIV*," followed by the page number.

[2] Alfred Jules Ayer, *Language, Truth and Logic* (London: Victor Gollancz Ltd., 1948), pp. 107–108. Hereafter cited in the text (in parentheses) as "*LTL*," followed by the page number.

influence others to have feelings similar to my own where X is concerned. And surely, any such attempt to so communicate or influence would appear to presuppose a judgement which Ayer's view does not permit. Can we admit as adequate a view which goes so far as to even disallow the possibility of expressing approval or disapproval via ethical statements? I think not.

Yet to point out the disallowance by Ayer of the possibility that we are voicing sentiments of approval or disapproval when we make an ethical claim is to do less than justice to the purely negative character of his position. "Sentences which contain normative ethical symbols are not equivalent to sentences which express psychological propositions, or indeed empirical propositions of any kind," he tells us (*LTL*, 105).[1] What is here being denied is that any translation of ethical to nonethical terms is consistent with the conventions of our language (*LTL*, 24).

For example, Ayer tells us that when we say to someone, 'you acted wrongly in stealing that money,' we are saying nothing more than if we had simply said, 'you stole that money,' coloring our statement with a peculiar tone of horror. In the light of the position outlined, such a reduction would appear to constitute a proposal to do away with the term 'wrong.' It is as if Ayer had said: "Let us abolish the word 'wrong' for we can better indicate our meaning by colouring our ordinary declarative utterances with a peculiar tone of horror."

Such a proposal seems plainly to overlook two obvious factors, either of which is sufficient to render it inadequate. Firstly, we might well ask how we could ever be sure that we were not mistaken in our interpretation of the emotional tone coloring the statement at issue. Ayer, I think, would have to admit that no guarantee that we were not mistaken could be offered. But if this is the case it were perhaps better to abandon ethics altogether than subscribe to an ethics at

[1] Ayer is careful to qualify this assertion by drawing a distinction between 'normative' and 'descriptive' ethical symbols. Descriptive symbols express ordinary sociological propositions. In justice to Ayer it must be admitted that the theory we are ultimately concerned to advocate would probably be classified by him as sociological and thus irrelevant. To which it can be replied that it is only irrelevant if Ayer's fourfold classification of ethical systems be accepted.

the mercy of individual nuance and inflection. Secondly, it seems sufficient argument against Ayer's proposal to point out that such common ethical terms as 'wrong' do possess an ordinary acknowledged use function; that this function is commonly seen to consist in, partially at least, the communication from one agent to another of a certain state of mind defined by the ethical term in question, and finally, that this common use satisfies human requirements in a superior and less ambiguous manner than Ayer's proposal does.

In the light of the foregoing criticisms, I conclude that, considered as an answer to the question with which we began our inquiry, Ayer's view represents one legitimate possibility, but that its adequacy as an answer is highly dubious.

III

> I think that the theory can be adversely criticised also on the grounds of his holding that logic and knowledge cannot help in the formation of *better* ultimate goals. It seems to me that ultimate goals can be formed and adopted either reasonably or unreasonably.[1]

The ethical view of Rudolf Carnap marks a midpoint between the positions held by A. J. Ayer and C. L. Stevenson. Ayer's view had disallowed to ethical propositions even the function of expressing subjective attitudes; Stevenson's view admits an element of belief into ethical statements and thus makes possible the judgment of them as true or false. Carnap takes normative ethical statements or value judgments to be expressions of imperatives. To voice an ethical statement is, for him, to utter a command. Ethical propositions cannot reflect beliefs, for value statements do not admit of truth or falsity. On the other hand, value statements "may have effects upon the actions of men" and thus can be held to reflect attitudes in the Stevensonian sense of the word.

On Carnap's view, normative ethical statements and value judgments are synonymous. But the characterizing feature of normative ethical statements is the imperative element in them. Such statements, directly or indirectly, reflect a com-

[1] Harold Newton Lee, "A Semantic Criticism of Reichenbach's Ethics," *Ethics*, LXIV (1953–54), 310.

mand in some form or other. If value judgments are synonymous with such statements, it follows that they, too, despite their assertive form, express imperatives. In Carnap's words: "a value statement is nothing else than a command in a misleading grammatical form."[1]

In speaking thus, Carnap achieves a position of extreme ethical simplicity, but it seems open to question as to whether the virtue of simplicity is sufficient compensation for the risk of excluding those value judgments which at first glance do not seem to fall under the category of commands. For clearly the only ethical statements of which it could be said that they obviously are commands are those in which one person says to another 'you ought to do so-and-so.' Whenever ethical statements have reference to any person other than the person directly addressed, or to the past in general or to an unfulfilled past condition, or to a future considered merely as possible, the command theory seems to violate ordinary usage in that if offers us what normally regard as a distorted account of what we actually say or hear. This is not to deny that there is an imperative element in many, if not all, of our ethical utterances. It is to deny that a theory which confines itself to imperatives is wholly adequate.

For example, it is difficult to see how such a statement as 'Socrates was a good man' could be phrased so as to bring out its imperative character. One might use Stevenson's formula and translate the statement to read 'I approve of Socrates; you do so as well.' This to be sure brings out clearly what imperative force the original statement implicitly possesses. But it also plainly illustrates a non-imperative component which may be interpreted either as an expression of an attitude or the voicing of a belief.

If what has been said above has any force, Carnap's view that ethics reduces to imperatives must be held to be not entirely adequate. But in justice to Carnap it must be questioned whether the normative ethical statements or value judgments he is here characterizing constitute for him the whole

[1] Rudolf Carnap, *Philosophy and Logical Syntax* (London: Kegan Paul, Trench, Trubner, 1935), p. 24. Hereafter cited in the text (in parentheses) as "*PLS*," followed by the page number.

of ethics, for if they do not then his view cannot so arbitrarily be dismissed. And in fact such normative ethical statements do not for Carnap constitute the sole definition of ethics. Of prime importance here is the qualifying adjective 'normative.' Ethics has two senses, two definitions. The one we have thus far considered is the sense of ethics as a philosophy of moral values or moral norms. Another, and perhaps more important sense, is the usage of 'ethics' as a descriptive term characterizing "psychological and sociological investigations about the actions of human beings, especially regarding the origins of these actions from feelings and volitions and their effects on other people. Ethics in this sense is an empirical, scientific investigation; it belongs to empirical science rather than to philosophy" (*PLS*, 23).

Obviously, ethical statements in this latter sense are not reducible to imperatives but have status as factual assertions about the conduct of the society they are referent to. Unfortunately, Carnap does not elaborate to the extent of providing examples illustrative of statements about ethics taken in this latter sense, However, he does tell us (*PLS*, 26) that historical and psychological propositions providing analysis or causal explanation of acts of valuation and value statements do constitute meaningful scientific propositions.

It is when he goes on to say that "the value statements themselves are here only objects of investigation; they are not propositions in these theories, and have, here as elsewhere, no theoretical sense" (*PLS*, 26), that we must take exception. Such a position seems to represent no advance beyond the command theory. Value judgments are rejected as having no theoretical sense. Those statements which do have theoretical sense are assigned to empirical science rather than to philosophy. In so far as the latter is concerned ethics is still reduced to the expression of imperatives. We thus end where we began.

I have argued that the characterization of value judgments as implicit commands is not wholly adequate or in keeping with common linguistic usage. I shall argue in Section Six of this paper that value judgments can meet the requirements for scientific hypotheses laid down by Carnap while yet retaining force as value judgements. Carnap's theory, like Ayer's before it, seems, to me, to represent one possible, though not wholly satisfactory, view.

IV

> The emotive theory of value is a recent aberration that also overemphasizes the affective factor in the contexture. Always to call the affective factor an emotion involves a peculiarly imprecise use of the term "emotion" ... The emotive theory is guilty not only of this semantic error, but also of the graver error of identifying the whole with a part. Synecdoche is a legitimate figure of speech, but philosophic theory should be more than a figure of speech (*MIV*, 188).

Ayer has sought to solve the ethical problem by the denial of any real function to ethical terms. Carnap has implicitly reduced ethics to the expression of imperatives. Neither philosopher has ventured on the detailed analysis that must precede any such summary rejection or reduction. With the publication of *Ethics and Language*[1] the emotive theory comes of age.

If we adhere to the principle that the meaning of a term is its use, it must be conceded that the constant common usage of such peculiarly ethical terms as 'good,' 'bad,' 'right,' and 'wrong' attests to the fact that they do have a meaning. What distinguishes Stevenson's work from that of his predecessors is (1) his realization that the everyday usage of common ethical terms requires a detailed analysis of their use function, and (2) the freshness of approach which he brings to the undertaking. Stevenson's view, characterized as it is by an insight and originality not to be found in the ethical writings of Carnap and Ayer, deserves and requires our careful consideration. For on the adequacy of this view the tenability of the emotive theory largely rests.

Of central concern to Stevenson is the answer to the

[1] I am aware that the ethical view advocated by Stevenson in his articles for *Mind* and other publications is somewhat different from that advocated in his book. However, since these various articles all antedate the book in order of appearance, and since the basic argument of many of them has been incorporated into the book, I am assuming that the view therein expressed represents Stevenson's most mature position.

question: What is the nature of ethical agreement and disagreement? The nature of the answer is indicated at the outset. Agreements and disagreements are classified into those expressing beliefs and those reflecting attitudes. Disagreements involving opposition or interpretation of facts are classified as disagreements in belief.[1] All factual disagreement on scientific issues falls under this classification. The importance of this last for ethics is obvious. For if ethics be regarded as a science, then all disagreement in ethics must be basically agreement or disagreement in belief. Yet ethical disagreement seems of a somewhat broader nature than this. We are faced with situations involving opposition of purposes, aspirations, wants, preferences, desires, etc. Disagreements of this nature Stevenson terms disagreements in attitude. All ethical disagreement then is analyzable into disagreement as to attitude or belief. But this is not to hold that these two classifications are independent of each other. Logically speaking, they may well be, but factually the two types interpenetrate in nearly every case. Thus we may have disagreement in belief about attitudes, or difference in attitudes towards beliefs shared in common. In Stevenson's words, "when ethical issues become controversial they involve disagreement that is of a *dual* nature. There is almost inevitably disagreement in belief ..., but there is also disagreement in attitude."[2]

Two questions must be faced with regard to this classification: (1) is it exhaustive? (2) is it adequate? As concerns the first question there would seem to be some doubt. Moralists often disagree about methods. Ethical theorists differ with respect to the usage of terms. Nevertheless, I do not feel that the elucidation of these additional types points to any serious inadequacy in Stevenson's classification. Disagreement about methods is largely disagreement about means to the attainment of specified ends, and if we analyze such disagreement in terms of attitudes and beliefs about ends, in all cases where

[1] With Stevenson, we shall hereafter confine our attention to disagreement, treating the positive term by implication.

[2] Charles L. Stevenson, *Ethics and Language* (New Haven: Yale University Press, 1944), p. 11. Hereafter cited in the text (in parentheses) as "*EL*," followed by the page number.

disagreements as to methods are not classifiable as special cases of disagreement in attitude, they certainly seem classifiable as special cases of disagreements in beliefs or beliefs about attitudes. And what has been said with regard to disagreement about methods seems equally applicable to disagreements about the correct usage of terms.

With regard to the exhaustiveness of Stevenson's classification, we may, I think, conclude that while other types of disagreement may be adduced, none is of such an independent nature as to preclude its subsumability under the two basic classifications here considered. And if this is so, then an affirmative answer to our second question seems to follow automatically. For that which is admittedly exhaustive as a classification must surely, as a classification, be regarded as adequate.

The particular variety of emotivist theory advocated by Stevenson takes its rise in the presupposition that the translation of ethical into non-ethical statements is not only possible but necessary. Such translation, if it is to be accepted as adequate, presupposes a more or less inflexible pattern or form of translation. On Stevenson's view the need for pattern or form is supplied by the "working-models."

Stevenson distinguishes two patterns, two working-models. In the first pattern, any declarative statement having a peculiarly ethical term as predicate is translatable into a statement of the form, 'I approve of X; do so as well' (*EL*, 21). In the second pattern, any statement of the form, 'A is good (or bad, etc.)' is translatable into a statement of the form, 'A has qualities or relations X, Y, Z . . .,' except that "good" (or bad, etc.) has as well a laudatory emotive meaning which permits it to express the speaker's approval and tends to evoke the approval of the hearer (*EL*, 207). Both patterns must be considered. However, since the greater portion of the analysis is carried on in terms of the first pattern working-model, and since Stevenson himself notes that "the importance of the second pattern lies not in revealing new ethical content but rather in revealing new complexities of language" (*EL*, 210), our discussion of working-models as such will be primarily directed at the analysis offered in terms of the first pattern.

Further, since the analysis in terms of the second pattern is clearly allied with Stevenson's doctrine of "persuasive definitions," we shall reserve our remarks on the second pattern analysis to our discussion of this latter doctrine.

The discussion of the first-pattern analysis falls naturally into two parts, corresponding to the two fundamental questions which must be raised in connection with any translation by means of the working-models. These questions are, (1) is such a translation at all possible? and, (2) assuming that it is, is it adequate?[1] The latter question first.

At the outset it must be recognized that the working-models under consideration are not precision instruments. In Stevenson's own words, they are somewhat "crude" (*EL*, 31). In all fairness to our author it must be here remarked that the distinctly critical treatment to which we propose to subject them is undertaken with this reservation in mind. Any final assessment of the adequacy of Stevenson's view must take this acknowledged crudity into account.

'I approve of virtue; do so as well.' In terms of the first-pattern working-model this is, in principle if not always in detail, an adequate translation of the statement 'virtue is good.' It is apparent, however, that any conclusion as to the adequacy of the translation presupposes the conviction that any ordinary ethical statement of the declarative form can be analyzed into an expression of personal approval or disapproval of the subject of the statement. Any decision with regard to the adequacy of the first-pattern analysis presupposes a decision as to the adequacy of the notions of approval and disapproval to express ethical judgments. Our first concern must be to determine the ethical significance and validity of these notions. For convenience sake we limit our discussion to the notion of approval.

To begin with, it may be contended that the word 'approval' is vague and ambiguous, and that therefore any translation which involves this notion cannot be held too precise. 'Ap-

[1] A third fundamental question is: is translation necessary? For the purposes of the present argument we shall assume that it is. However, when later we come to the consideration of possible alternatives to the Emotivist view, this third question will assume prime importance.

proval' surely seems to cover a wide range of feelings. Extensionally taken, it is vague. Taken intensionally, it is often ambiguous. As Ayer has pointed out (*LTL*, 104), we approve of many things to which we do not ascribe the character of moral worth. Moreover, it seems at least questionable as to whether we can distinguish a specifically moral variety of approval from approval in general without having recourse to metaphysical presuppositions unallowable on any variant of the emotive theory. And all that has been said with regard to the word 'approval' seems equally applicable to the term 'good.' If we confess to the ambiguity and vagueness inherent in ethical terms, as Stevenson has (*EL*, 34), and in addition must admit the difficulty of rendering 'approval' precise, can we really claim to have provided a thoroughly adequate translation when we assert that 'virtue is good' is equivalent to 'I approve of virtue; do so as well?' I do not think that we can, and apparently neither does Stevenson, for at the close of his chapter on the working-models he has this to say:

> They (the working-models) provide a meaning which can be assigned to ethical terms, and which is well suited to certain contexts. Yet if they are taken as typical of all contexts, and sufficient in themselves to clarify common usage, they will represent only another contribution to linguistic fiction (*EL*, 36).

The adequacy of the working-models, even in the view of their author, seems limited. Precisely how limited, or how inadequate, can only be ascertained by further examination of the notion.

When Stevenson remarks that "moral judgments are concerned with recommending something for approval or disapproval" (*EL*, 13), we can surely agree. But, we must immediately add, in most cases this does not seem to be all that they are concerned with .The adequacy of 'approval' as a translation term seems questionable on the grounds that the expression of approval or disapproval does not appear to exhaust one's meaning or intent when one voices a declarative ethical statement.

For in voicing an ethical statement one may, in Stevenson's own terms, be expressing a belief as well as conveying an attitude. For example: I might express my conviction that knowledge is good. In so doing, I indicate an attitude towards knowledge, but I also imply a belief about the intrinsic value of knowledge. I might render this belief more explicit if I rephrased my statement and said 'I believe that knowledge is good.' If, as Stevenson has admitted (*EL*, 13), the emphasis on the notion of approval sanctions only the expression of attitudes and gives no emphasis to the implicit element of belief involved in many ethical statements; if a further investigation of the imperative component of the working-model is necessary in order to elicit this element of belief, then, taken at its face value, 'approval' seems to lack the ability to convey the fullness of the meaning implicit in the judgment it is used to translate.

Much of the force of the above criticism appears to evaporate when we have regard to the imperative component of the working-model. It is this part of the translation statement 'do so as well' which implicitly reflects the element of belief. In providing reasons why others should heed my imperatives, I reveal the beliefs that underlie my ethical assertions. Now it is true that many ethical statements have a meaning that is in part imperative. Many others, including all statements of the form 'you ought (or ought not) to do X,' seem almost wholly imperative in tone. But must all ethical statements of the form 'X is good (or bad etc.)' possess an imperative component? If a negative answer to this question is possible, it would appear that in any case where an imperative component is not to be found the working-model analysis could not apply. For in such a case the statement would express a judgment that either would not admit of translation in terms of 'approval,' or, if it did, would imply a concept of approval signifying much more than the customary connotation of the term.

When I declare my conviction that 'virtue is good,' I am not necessarily demanding assent to that conviction. The assent of others would please me; support for my conviction from other quarters would strengthen my belief in the soundness of my belief. But lack of such support would not necessarily cause me to abandon that belief. In short, while my ethical utterances may implicitly influence others to the adoption of my point of view, it is not always the case that I am explicitly concerned to gain the approval of others for my ethical convictions. And to the degree that I am not, the imperative component of the working-model loses relevance.

That I am not always concerned to gain the assent of others to my ethical convictions is perhaps best illustrated by reference to the element of belief underlying any ethical statement. That some such element of belief is always present becomes clear when we analyze the implications of the term 'approval.' 'Approval' invariably presupposes some insight into that which is being approved. That is to say, 'approval' inevitably demands a reason for the approval. And if it be conceded that there is invariably a reason to be found for that approval, then it would seem that this reason must involve reference to the belief or beliefs which underlie it. The concept of approval itself appears to contain an implicit if not explicit reference to beliefs in addition to its explicit reference to attitudes.

But if an ethical statement is expressive of a belief, it is either true or false. The making of a statement possessing truth value does not necessarily require the assent of others to its truth, for it is true whether other agree or are indifferent.

Yet if this is so, then our need for an imperative component seems to have virtually vanished. The declarative component 'I approve of X', since it is seen to have implicit reference to beliefs in addition to its explicit reference to attitudes, stands complete in itself. The eliciting of an imperative component does not seem necessary in all cases to the translation of an ethical statement.

In sum, with regard to the notion of approval, it must be recognized at the outset that the term is vague and ambiguous, that it is employed by Stevenson as a translation term for a word (good) which itself is vague and ambiguous, and that in consequence, the resultant translation can be at best imprecise and at worst erroneous. Further, it is open to question whether or not the use of 'approval' as a translation term exhausts the meaning of the translated term 'good.' While the original statement does undoubtedly imply the notion of approval, it does more than that; it evinces a belief which the term 'approval,' in so far as it is taken to signify attitudes, fails to express.

The expression of beliefs is, on the theory here under consideration, supposed to be elicited by an analysis of the reasons for adherence to the imperative component of the translation sentence. But it does not seem to be the case that an imperative element is to be found in all ethical statements since these statements often express beliefs whose truth or falsity may be tested by the utilitarian criterion. In other words, it does not always happen that assent is required to the expressed ethical statement, and to the degree that it is not, the ethical statement may be said to lack an imperative component.

Thus, on the one hand we are led to doubt the adequacy of the term 'approval' to express the declarative meaning of the ethical statement involved, and on the other we are led to question that the statement need always imply an imperative meaning. Both components of the working-model seem open to question as to their adequacy or necessity. However, this is not to deny the value or usefulness of the working-model analysis in general. For the working-models are soundly grounded in the recognition that most, if not all, value judgments imply the notion of approval and demand the assent of others to this notion. In most cases they (the working-models) are adequate to the task assigned them. Our demurrer is based upon the fact that they are not *invariably* adequate to this task, and such demurrance I feel certain Stevenson would be the first to recognize. We are not here concerned with questions of the rightnees or wrongness of ethical outlook but with its comprehensiveness. Our basic criticism resolves into the contention that the working-models do not provide a sufficiently comprehensive analysis of value judgments. Is there a possibility that they might be amended to do so?

If Stevenson's working-models are inadequate as they stand, do possible improvements in them seem to augur any prospect of greater satisfaction? For example; we might say that 'X is good' means 'I believe X is better than Y; do so as well.' The substitution of 'believe' for 'approve' seems to allow for the increased possibility of testing the truth or falsity inherent in the statement. However, it is doubtful whether we have really gained by the substitution, for such a translation is useless unless we are in prior agreement on the value of the comparative 'Y.' I might say 'it is more blessed (or better) to give than to receive,' but unless the person addressed is aware of just what value 'receiving' has he has no way of evaluating my statement. If he agrees that to receive is good, or in Stevenson's language, 'I approve of receiving; do so as well,' is he helped any by my saying 'it is better to give than to receive?' I think not. At most he can say 'I approve of giving more than I approve of receiving,' but 'more' like 'better' requires a fixed standard by which to measure, and such a fixed standard is nowhere allowed for in Stevenson's system.

A key problem for ethical theory is that of deciding whether or not there are things or acts that are intrinsically good or bad, self-evidently right or wrong as the case may be. For Stevenson the problem becomes one of determining whether the first-pattern working-model can sanction an emphasis on 'intrinsic value,' and whether it can recognize a practicable division of labor between those who recommend ends and those who study means.[1]

Thus, in terms of the working-model, 'I approve of X intrinsically' translates into 'I approve of X when I disregard all of its consequences upon other objects of my attitudes' (*EL*, 177). It must be kept in mind that Stevenson is here assuming that the statement 'X is intrinsically good' is translatable into 'I approve of X intrinsically.' As before, the question of the adequacy of the translation is in order. We need, I think, have no quarrel with that part of it which asserts 'intrinsic' to involve disregard for all consequences. However, when Stevenson speaks of "objects of my attitudes," I feel that he is on unsure ground. To assert that anything is good intrinsically seems to me to clearly involve more than the expression of an attitude. To say 'this is good in itself' is (a) to make a dogmatic statement, it is (b) to

[1] I confine the discussion to 'intrinsic value' since I think most ethical theorists would take the fact of 'extrinsic value' for granted.

express a conviction, (c) a belief. It might be better to speak of 'other objects about which I have beliefs and which give rise to attitudes' rather than to speak merely of 'objects of my attitudes.' But if we correct the translation in this manner we seem led back again to the objection that 'I approve of X intrinsically' is not an adequate translation of 'X is good intrinsically.' If the statement involves an assertion of belief, it seems to say something more than that the speaker is acting emotively to persuade his auditor to similar intrinsic approval of X. The translation would be adequate if we could characterize the statement 'X is intrinsically good' merely as an expression of its speaker's attitude, but this we cannot do. In making the statement we undoubtedly do attempt to influence the hearer to have an intrinsic attitude, but we seem in addition to be attempting to sway his beliefs. "People do not agree on the intrinsic value of something unless both approve of it intrinsically," remarks Stevenson (*EL*, 178), and I agree. But while approval may be a necessary factor it is by no means a sufficient one. Agreement on intrinsic value, assuming for the moment that there is such a thing, presupposes agreement as to convictions and in beliefs, as well as the possession of similar attitudes.

Having redefined 'intrinsic' in terms of his working-model, Stevenson further proposes that to say 'X is intrinsically good' is the same thing as to say 'X is good as an end' (*EL*, 178–9). Proceeding from this assumption the remainder of his analysis of intrinsic and extrinsic value is conducted in terms of ends and means. It seems pertinent to ask whether such an interchange of terms is warranted. Is 'intrinsically good' synonymous with 'good as an end?' I suggest that it is not. 'Good as an end' taken strictly, seems to imply that the end referred to is good only when desired, and therefore only when nonexistent. For that which is desired as an end lies in the future rather than the present. On the other hand, it seems plain that when we speak of something as 'intrinsically good' we mean to imply its present existence in some mode. Further, that which is declared to be 'intrinsically good' is averred to be so at all times and under all conditions. Desire is irrelevant to such an assertion. Here, as in the case of the translation of ethical statements in accord with the first-pattern working-model, Stevenson's assumptions as to the synonymity of terms and phrases appear questionable.

Even so, it may be justly said that we have still come to no conclusion as to whether or not there is such a thing as intrinsic value. Consideration of Stevenson's answer to this question is still in order. His answer is simple and direct. "Intrinsic attitudes are not an immutable part of each man's nature," he tells us (*EL*, 192). Moreover, he declares that "among the factors which determine intrinsic attitudes we must recognize habituation" (*EL*, 193). He nowhere, as far as I have been able to find, tells us what the other factors that determine such attitudes are. We are likewise left uninformed as to whether "habituation" is responsible for the determination of beliefs as well as attitudes. If it be fair to tax Stevenson with the view that notions as to what is intrinsically good are formed by environment, then on his view we should conclude that such recognized intrinsic goods as knowledge are so recognized simply because we have in some way through process of habit come to see their intrinsic value.

Yet here again we meet a stumbling block. 'Intrinsic' connotes that which is fundamental to the thing to which it is ascribed. If we admit that we can come to the awareness of an intrinsic value in things, we seem to be admitting our faith in an objective good independent of our attitude towards it. But to so conclude would be foreign to the letter as well as the spirit of Stevenson's doctrine. Faced with such a conclusion we might well wonder if a doctrine such as Stevenson's can sanction any emphasis on intrinsic value at all. In other words, I would, following Harold N. Lee, suggest that the whole problem of intrinsic value is one that cannot be adequately dealt with on emotivist grounds. Our difficulties in the employment of the working-models, and the questionability of talking in terms of means and ends is clear evidence of this.

The first pattern of analysis has shown certain deficiencies as an instrument for eliciting the meaning of ethical statements. We must now consider whether the second pattern provides any more precise instrument for elucidating their significance. At first glance it seems not. Stevenson prefaces his discussion with the admission that "in both patterns there is an emphasis on disagreement in attitude. The distinguishing features of the second pattern, then, lie solely in the added descriptive meaning that it provides" (*EL*, 206). "The importance of the second pattern lies not in revealing new ethical content but rather in revealing new complexities of languages ... it is chiefly of interest in calling attention to a much used but too little studied type of definition" (*EL*, 210). This type of definition is what Stevenson calls "persuasive definition." A persuasive definition is one which purports to alter the descriptive meaning of a term whose previous meaning is both descriptive and strongly emotive (*EL*, 210). This alteration is accomplished by giving the term "greater precision within the boundaries of its customary vagueness; but the definition does not make any substantial change in the term's emotive meaning" (*EL*, 210).

The aim of persuasive definition is to plead a cause. To say 'X is good' is to laud X in addition to expressing mere approval of it. The second pattern of analysis permits of a wider range of attitudes towards that thing or act characterized by the ethical term, but does not seem designed, nor does it function, to overcome the translation deficiencies of the first pattern working-model.

We have throughout been primarily concerned to determine the adequacy of the working-model analysis to convey the meaning implicit in value judgment. It is only when the meaning of the ethical statement is ascertained that the question of the resolvability of ethical disagreement can be dealt with. What we ultimately require of Stevenson, however, is a method adequate to the solution of ethical disagreement. Yet such disagreement only seems resolvable if the statements involved admit of being judged true or false. Such judgments involve rationalization, and Stevenson concedes that "rational methods alter attitudes only through altering beliefs" (*EL*, 234). "Rational methods can resolve ethical disagreement if and only if it is rooted in disagreement in belief" (*EL*, 237). I fully agree. To the extent to which the working-models of the first and second patterns stress disagreement in attitude rather than disagreement in belief, they are of small help in either resolving or clarifying ethical disagreement. In concluding thus, I am aware that Stevenson would reply that it is not his task or aim to provide techniques for the resolution of ethical problems. His concern is with the methodology of ethics rather than with the solution of the problems it involves (*EL*, 1). Nevertheless, to suggest a new approach, to set up and argue for the employment of criteria, to apply new techniques to ethical statements, and to draw conclusions from their employment, is to dabble in questions of normative ethics whether one admits it or not.

The point is simply this; the non-philosopher, when he voices a value judgment, does so in the firm belief that his utterance is both significant and conformative to ordinary usage. This belief may have its foundation solely in introspection, but while introspection cannot be claimed to be an infallible guide, neither can it be arbitrarily dismissed as of no consequence as Emotivists are occasionally prone to do. To claim that in making ethical statements the non-philosopher is merely expressing an emotion or indicating an attitude is to claim that he is violating ordinary usage, or that if he is not, then it is to claim that ordinary usage is in these instances misleading.

If we elect the first alternative we seem to be voicing a contradiction, for what we assert is that the ordinary usage of mankind is not ordinary usage at all when ethical statements are involved. This seems a bit unjustified. If we elect the second alternative how are we ever to justify the meaning of the principle 'the meaning of a term is its use?' If ordinary usage is misleading, what is left as our guide? A third alternative is possible. We may say to the common man that he is not violating ordinary usage, but that his statement has no empirical significance. But if 'empirical significance' is partially grounded in 'use,' if that which has a use is empirically significant, I fear that the emotive theory will leave our common man scratching his head in puzzlement and muttering imprecations against all philosophers, positivists or otherwise.

It is a strange fact that nearly all ethical theorists bring up the question of ordinary usage as an argument against theories opposed to their own. It is a standard gambit among writers on ethics to assert that in some degree their opponents depart from common sense views of 'good,' 'right,' 'ought,' etc., or, in other words, that they define ethical terms in such wise as to violate our common sense view of how these terms should be employed in reference to things or acts. Speaking in Wittgensteinian terms we might say that philosophers who hold an emotive theory of ethics are merely stressing the resemblances of ethical propositions to expressions of emotive approval. Philosophers who dissent from the emotive view are simply stressing the differences between emotive states of approval and disapproval and ethical propositions, and neither side seems of much aid in resolving ethical problems.

Finally, we might remark that all varieties of the emotive theory seem to arise not from a consideration of the ethics possible presupposing Wittgensteinian principles as given, but rather from a need to justify ethics in the light of their empiricist or naturalist epistemologies. This is certainly so in the case of Ayer and Carnap, obviously much less so in the case of Stevenson. Yet even granting that Stevenson's point of view is the result of an attempt to ground ethics in Wittgensteinian principles rather than an effort to justify an epistemology, the difficulties which seem inherent in any position which reduces ethics to emotive states seems to demand that other avenues be explored. It is not necessary on Wittgensteinian principles to hold to an emotive theory of ethics. Let us consider some alternatives.

V

> The present essay maintains that values can be shown to be part of a factual context, and in consequence, the contrast between facts and values is not well taken. Thus, the conclusion that statements about moral values are no more than volitional or emotive utterances based on the attitudes and infected with the non-propositional nature of the attitudes is not well taken either. The conclusion that moral statements are only about attitudes is based on an incomplete analysis of the total complex situation in which moral values arise (*WMJ*, 89).

The relegation of ethics to the realm of the emotional or psychological is based on the assumption that all non-empirical statements are analytic, and that all ethical statements are non-empirical. Granted this, and proceeding on the assumption that the only significant statements are synthetic and empirical, i.e., verifiable in experience, it seems to follow that there can be no significant ethical statements.

I wish to contend that neither of the above assumptions is necessarily the case. It does not appear to me to be necessary to concede that all non-empirical statements are analytic. 'Virtue is good' would, I think, be classified by Wittgensteinians as a non-empirical statement. Is it analytic? Only, I suppose, in those cases where 'good' is defined as virtue, but even in ethical history such cases are not many. Moreover, is it always the case that all ethical statements are non-empirical? I say knowledge is good and define 'good' as that which is useful to mankind in general, that which has utility, that which is of benefit or pleasure. Is then the statement that knowledge is good non-empirical? I cannot see that it is.

If the good is the useful, then all I need to ascertain in order to verify the statement is that knowledge has some use. This I do by empirical reference. My statement is strictly verifiable. Appeal to the world of fact is possible to determine its truth or falsity. I am aware that the making of such a statement, involving as it does a multitude of fundamental epistemological assumptions, may appear somewhat arbitrary at first glance. The remainder of the argument of this paper has as its intention the clarification of the assumptions, and the exposition of the position, on which it is based.

The advocate in Positivist ranks of the ethical theory that the good and the right is, ultimately, that which brings pleasure or has utilitarian value is Moritz Schlick. His theory of ethics represents a departure from the emotive theory that yet, in a sense, is no departure, "Ethics has to do entirely with the actual."[1] "Its basis is psychological" (*PE*, 28). The question we must ask ourselves is why are things good or bad, right or wrong. "The central problem of ethics concerns the causal explanation of moral behavior; all others in relation to it sink to the level of preliminary or subordinate questions" (*PE*, 28). So far there seems nothing that any Emotivist could seriously differ with. It is only when Schlick asserts that "ethics has to do only with resolutions" (*PE*, 32), when he inquires into the motivations of moral actions, that the departure becomes evident. There is little or no talk of 'motives' or 'will' to be found in Stevenson, Carnap and Ayer; it is central to the ethics of Schlick.

"Every experience has an emotional tone that is pleasant or unpleasant, ... in every experience there is a feeling of pleasure or pain" (*PE*, 37). Moral demands or duties are, in the last analysis, traceable back to the feelings of pleasure or pain of the individuals composing the society. Moral commands arise out of human needs and desires. "In the moral valuation of definite acts or dispositions as good or bad is mirrored only the measure of joy or sorrow that society expects to receive from those acts or dispositions" (*PE*, 85).

Ethics is grounded in psychology, but it is not purely emotive in character. "No matter what one's conduct is one always follows some inclination" (*PE*, 62). And this being so, one has a hedonistic basis for ethical judgments. "The law of the will asserts that one can act only towards that end whose idea is most pleasant for one, and this means in the direction of the strongest impulse" (*PE*, 98). This appears to be a straightforward doctrine of psychological hedonism, psychological hedonism usually being defined as that doctrine which holds that we are so constituted psychically as to

[1] Moritz Schlick, *Problems of Ethics* (New York: Prentice-Hall, 1939), p. 21. Hereafter cited in the text (in parentheses) as "*PE*," followed by the page number.

pursue always our own pleasure. Yet Schlick is more than a psychological hedonist; he is an ethical hedonist as well insofar as he adopts the principle 'be ready for happiness at all times' as his maxim. Moreover, he is also a Utilitarian insofar as he regards the idea of a happy community as a pleasant idea (*PE*, 98).

This sketchy summary of Schlick's view has had as its main purpose the buttressing of the view that other than emotive theories of ethics are possible on Wittgensteinian principles. We do not, however, propose to subscribe to such a theory as the one outlined above. Hedonism in general, and psychological hedonism in particular, has long served as a whipping boy for moralists ancient and modern. The arguments refuting it are many, stereotyped, and well-known.

If its (hedonism's) basic assumption be that only states or activities of the agent himself can be the object of pleasure, then it seems either analytic or, if synthetic, refuted out of hand by reflection upon the motives from which we actually act. As a general objection I do not think that the tenor of Schlick's hedonism is such as to evade this criticism. However, for our purposes it will be sufficient to test the tenability of Schlick's system in the single case on which he lays greatest stress.

For Schlick, of the ideas which function as motives, that one gains the upper hand which finally possesses the highest degree of pleasant emotional tone, and thus the act in question is unambiguously determined (*PE*, 39). The law of motivation may thus be expressed: In the conflict of motives the decision goes to the most pleasant or the least unpleasant (*PE*, 41). Even if we have regard to Schlick's statement that we must not look upon the genesis of acts of will as a static balancing but rather as a dynamic process (*PE*, 40), his conception of decision does not seem to hold. We surely do at times deliberately choose the least pleasant alternative in relation to both long and short run consequences.

Consider the case of the martyr. On Schlick's theory, the martyr's action is explicable solely on the grounds that such action leads to the most pleasant state. If we have regard to physical consequences it seems a gross misusage of language to call martyrdom a pleasant state. If we have regard to mental states it is possible to conclude otherwise. One can argue, and Schlick does, that the martyr is obsessed by a cause. The thought of the particular goal or idea possessing him dominates his consciousness to such an extent that there is little room for anything else. It may be true that the idea of painful destruction can intrude upon this thought, but the inspiration that the thought provides more than offsets any thought of pain. In Schlick's words, "inspiration is the greatest pleasure that can fall to the lot of man. To be inspired by something means to be overcome by the greatest joy in the thought of it" (*PE*, 46). I agree absolutely. The sole flaw in the argument is that it simply is not the case that all acts of martyrdom are endured under the emotive state of inspiration. Martyrdom may, and has been, endured with the full and intense awareness of the pain involved, and without any inspiration to drive away the insistence of the pain. Martyrdom is, in most cases, suffered as the result of a refusal to give up convictions of reason or faith. Belief in the rightness of a cause, or the stubborn insistence on maintaining the truth at all costs, cannot be analyzed into inspiration. Socrates did not drink the hemlock because he was inspired, but because of a conviction that truth was more important than life. Sydney Carton in Dickens' *Tale of Two Cities* died willingly not because he was inspired but because of a conviction that it was his duty to make amends for a worthless life by preserving worthier lives. Many instances might be cited; one counter-instance is sufficient to cast doubt on the adequacy of Schlick's theory.

However, it is important to realize just what it is we are refuting. Schlick's psychological hedonism is grounded in the assertion that even in the extremest case (martyrdom) the motive is still pleasure because the emotion of inspiration is the most joyful of emotions. We do not deny that inspiration is emotively grounded, nor do we deny that conviction or rational stubbornness or a sense of duty, or any other motives that may lead one to martyrdom, may have their roots in emotive states. It may well be, as Westermarck holds, that there are moral emotions although the difficulties of distinguishing them are many. All that is here denied is that these so-called moral emotions are necessarily pleasant emotions. If they are not, then we can say that when one acts from any of them, rather than from inspiration, the pleasure balance of one's action is negative, and the assertion that in the conflict of motives the decision always goes to the most pleasant or the least unpleasant is proven erroneous.

We must reject Schlick's view as inadequate, at the same time recognizing its worth for our analysis. What Schlick has shown is that ethical statements may, as such, meet the requirements for significant sentences without having recourse to a strictly emotive theory. For Schlick what is good or moral is that which conduces to the pleasure of the individual or the society. We cannot quite correctly say on this view that the good is pleasure, but we can say that those things or actions which have a surplus of pleasure value over pain value are good. Therefore if we say 'X is good,' if 'good' be defined in Schlick's sense, then the statement admits of verification in the empirical sense, for all that is necessary to verify it is to determine whether or not the thing or act under consideration actually does conduce to pleasant feeling on the part of the subject or subjects to which it is relative. This argument will probably not satisfy those who are in search of strict definitions. But if we can give the exact conditions under which ethical terms are applicable, need we seriously concern ourselves with grammatic definition? I think that we may well agree with Schlick on this point when he remarks that "if ethics ended with a definition it would be at most the introduction to a science" (*PE*, 6).

The importance of a view such as Schlick's, in my opinion, lies in the fact that it points the way to the possibility of a non-emotivist ethics, one that can be in accord with Wittgensteinian principles, yet one which avoids the necessity of violating ordinary usage to the extent of asserting that when we utter value judgments we are not voicing any statement of significance, but are at worst making a factual statement with emotional stress, and at best expressing an attitude of approval or disapproval which may or may not possess moral significance. Schlick has pointed the way, but he has not shown that his criterion of verifiability, pleasure, is an adequate or consistent one. Our task is to advance a criterion which in all instances fulfills this requirement.

VI

> Value, in the broadest sense, is a potentiality in an object (including those potentialities that become actual) ... Potentialities are what they are only in the light of actualities. Actual intrinsic value is the property of an object whereby it affords pleasantness or satisfaction in a complete contexture ... (*MIV*, 195–6).

Wittgenstein preaches the identity of grammar and use. The grammar of an expression includes its use. Meaning and grammar are not separate aspects of a statement. The significance of this thesis for ethics appears to consist in the idea that if we can show that ethical statements are functional, that they serve a purpose, we have made a place for ethics among the practical sciences.

In his article "Defending Common Sense,"[1] Norman Malcolm advances three criteria for ascertaining the correct use of language in regard to declarative statements. (1) There is a doubt to be removed. (2) A reason can be given for the assertion. (3) There is an investigation which if carried out would settle the issue. I suggest that on the view I shall put forth ethical statements such as 'X is good' or 'Y is wrong' satisfy all three of these requirements.

One person asserts that 'X is good.' This is certainly not a self-evident proposition no matter what X represents. The person to whom the statement is addressed may assent immediately, but it is more probable that he will seek by further questioning to elicit the reasons for such an assertion. I submit that the reason found to be involved in all such ethical assertions is the belief that the thing or act of which the ethical term is predicated will on examination be found to have a positive use value. By 'positive use value' I mean that when the various utilities or disutilities to society of certain acts or things are weighed against each other a positive

[1] Norman Malcolm, "Defending Common Sense," *Philosophical Review*, LVIII (1949), 201–220.

utility results. We say of an action that it is right. Examine the reason for such a statement and you will elicit a belief that the action is such as to conduce to the welfare of society on the whole, or to the benefit of some group or individual. Attention to the interests of the society, group, or individual benefited by the action labelled 'right' or 'good' constitutes the satisfaction of Malcolm's third criterion; the fact that the statement was not self-evident having satisfied the first criterion, and the reason advanced for its assertion having satisfied the second.

In accord with most utilitarians, I begin by assuming the primacy of good over right. I would contend that the right is determined only by reference to the good. Right action is action that is useful to society. The good is that which has a positive use value. But when we say this we must mean to assert more than that things or acts are useful in a material sense at the present moment. The utility of ethical statements about past and future actions must be affirmed; the utility of ethical praise itself must be accounted for. In the words of Henry Sidgwick,

> Although, in the view of a utilitarian, only the useful is praiseworthy, he is not bound to maintain that it is necessarily worthy of praise in proportion as it is useful. From a utilitarian point of view ... we must mean by calling a quality 'deserving of praise' that it is expedient to praise it with a view to its future production: accordingly, in distributing our praise of human qualities on utilitarian principles, we have to consider primarily not the usefulness of the quality but the usefulness of the praise: and it is obviously not expedient to encourage by praise qualities which are likely to be found in excess rather than in defect.[1]

The reference to expediency in the latter part of the quotation serves to point up the necessity of providing a brake on the indiscriminate application of 'good' to any and every thing or act. The doctrine of proportionality maintained in the early part of the quotation serves to narrow the application of 'good' to those things or acts more specifically believed to be 'morally good.' Nevertheless, it must be con-

[1] Henry Sidgwick, *The Methods of Ethics*, 5th ed. (London: Macmillan, 1893), p. 428.

ceded that to define good in terms of use-value is to employ the term in a somewhat wider context than would be allowed by many moralists. Unfortunately, this broadening of ethical terms seems to be the price we have to pay for obtaining inclusion for them within the realm of significant sentences.

Stevenson had proposed that ethical statements could be reduced to statements of fact, and thus rendered significant in the Wittgensteinian sense of the term, by translating them in accordance with his working models. I suggest that ethical statements can be rendered significant without translation or reduction to other forms simply by giving a utilitarian definition of the ethical terms involved. I submit that such procedure recommends itself in that by following it, the difficulties encountered in the employment of the working-models are avoided.

For example, to assert that 'truth-telling is intrinsically good and yet I do not always approve of it' is simply to make a factual statement to the effect that truth-telling is, as a general thing, useful or beneficial to society on the whole, but that in certain instances its use-value is negative owing to the harm resultant from it. To judge a law good and yet to disapprove of it, on the utilitarian system, is to say that the law serves a useful purpose and benefits society, but to express a personal distaste for it in particular either because it is not personally beneficial or because it is positively restrictive.

Similarly, with regard to comparatives, to say that 'X is better than Y' is to say that 'X is more useful than Y.' To say that it is more blessed to give than to receive is to say that a spirit of giving, as opposed to a spirit of taking, is more in accord with the useful aim of achieving the harmony and welfare of the greatest number, it being remembered, of course, that in all these examples the truth or falsity of the statement is determinable by an appeal to the experience of the society as to what constitutes the 'useful' for it.

Utilitarianism in its various forms is the outgrowth of the conviction that such terms as 'good,' 'right,' 'bad,' and 'wrong' should reflect the use-value of the objects or actions they are predicated of. To take as one's maxim the principle of 'the greatest good for the greatest number,' or to a dopt it in a form less vulnerable to criticism and say that we should term those things good and those acts right which conduce to the production of the greatest value to the society as a whole, is to express a belief that that which merits approval is the useful, and that which merits disapproval is that which in no way serves to further the interests of society. Schlick talks in terms of a surplus of pleasure value, and defines the good as that which is pleasurable. I suggest the replacement of 'pleasure-value' by 'use-value' on the grounds that the latter is far less vulnerable to criticism as an ethical criterion than the former.

It may be objected that such a proposal is of no help to ethics. We may say that fertilizer is good for the soil, or that John is good at games, or finally, that Gandhi is a good man. Only the last of these statements seems to bear any explicit ethical meaning; yet on our definition of 'good' all these assertions can be verified by pointing out the use-value to the individual or to the society as a whole of the subject of each statement. To argue thus, however, is merely to point out once more that ethical terms can have non-ethical applications and non-ethical meanings. The important thing to note about all of the examples given above, however, is that they all seem to satisfy the criteria for significant sentences. We do not arrive at the decision that 'Gandhi is a good man' is a significant sentence merely by marking its syntactical similarity to the other, non-ethical sentences; we assert its significance because we believe that it, like these, is open to acceptance or rejection as true or false. Its verification consists simply in pointing out that its subject was an asset to the society of which he was a member. Similarly, when I assert that 'helping old ladies across streets is the right thing to do,' I am making an ethical assertion. I am ascribing value to an action, and to one who denies that I am uttering a significant sentence, I reply that the obvious usefulness of such an action, its positive benefit to the lady in question, its negative benefit in that it may help prevent an accident, constitute ample verification of its significance.

Those familiar with Stevenson's analysis may at this point remark that the above explanation is merely a reflection of his couched in utilitarian language. What I am actually doing when I make the statement 'helping old ladies across streets is the right thing to do' is to express an attitude towards a specific action, and when pressed to justify my attitude I produce reasons for it that reveal my beliefs about the matter. To which I can but say again what has been said earlier, that while it is true that the statement has emotive value and tone, and undoubtedly indicates an attitude, it does more, it expresses a conviction or a belief, and it does this explicitly. The statement does not have to be translated in terms of any working-model in order for this belief to be brought to light. To say this is not to deny for a moment the probability that our use of such words as 'good' or 'right' always involves the existence of a favorable attitude in ourselves towards the object.[1] What we *express* when we call an object good or an act right is our attitude towards it, but what we *mean* is something about the object itself and not about our attitude towards it. Stevenson's analysis, focusing as it does on the 'expression', cannot help but conclude that what is primarily relevant is 'attitude', but if we have regard to 'meaning' we can see that what is not only primarily relevant but is of much greater importance for ethics is 'belief.'

To assert such is to assert the independence of ethics as a practical science. I say "practical" because on this view only those ethical statements which have a utilitarian application are considered valid. The statements of a purely theoretical ethics, like those of theoretical metaphysics, must be disallowed practical relevance. It makes no sense to say, for example, that the One is good or to speak of the moral autonomy of the will. Any ethical utterance which presupposes the acceptance of an ontological position must be disavowed, for ontology itself has no empirical status.

Yet, if on the one hand we disassociate ethics and ontology, on the other we must similarly disassociate ethics and epistemology, at least to the degree that ethics is dependent on any

[1] This seems to contradict our previous assertion that it is possible to say that something is good and yet to disapprove of it. Actually, however, no contradiction is involved because in asserting that something is good we *mean* we believe it.

given epistemological position. In a sense, ethics is dependent. In the sense that is, that all science or the acceptance of the reality of the world about us presupposes the acceptance of certain epistemological assumptions. But ethics is dependent only in this sense, and no more. It cannot be reduced to an offshoot of psychology since it is concerned with use-value and use-value can be empirically determined. It is not confined to the expression of emotive states or personal attitudes, for ethical utterances have meaning, have a use, serve a function. To assert a value judgment is to express a conviction that the thing or act judged is of a positive benefit to the individual or to mankind in general. What we need to know, and what ethics can tell us, is what is good for us and what is right for us. The empirical experience of the bad and the wrong is the ultimate witness of the relevance and correctness of all judgments of value.

Examination of the various positivistic paths to value has led us to fix on one as most likely to satisfy both Wittgensteinian principles and the requirements of common sense. It perhaps would be more apt to say that the position here contended for has been fixed upon as least susceptible to easy refutation rather than because it is most likely to merit approval. A comprehensive appraisal of its merits or defects would require a separate paper. Similarly, a decisive refutation of any of the alternative theories here cursorily considered would require a far more exhaustive examination than has here been attempted. The citing of one or two objections to any theory, damaging though they may be, does not provide an adequate basis for rejecting the theory outright. Theories must ultimately be judged as wholes. The inadequacy of any particular part, while regrettable, is not decisive. The merits of emotive theories as opposed to naturalistic or utilitarian theories are finally decidable only at the bar of common sense.

PUBLICATIONS BY HAROLD N. LEE

Books

PERCEPTION AND AESTHETIC VALUE (New York: Prentice Hall, 1938; reissued New York: Johnson Reprint Corporation, 1967).

Co-author and editor, ESSAYS ON THE THEORY OF VALUE AND VALUATION (Southwestern Philosophical Society, 1945).

Co-author, VALUE, A COOPERATIVE INQUIRY, edited by Ray Lepley (New York: Columbia University Press, 1949).

Co-author, THE LANGUAGE OF VALUE, edited by Ray Lepley (New York: Columbia University Press, 1957).

SYMBOLIC LOGIC (New York: Random House, 1961; London: Routledge & Kegan Paul, 1962).

Articles

"Morals, Morality and Ethics," *Ethics*, XXXVIII (1928).
"The Differentia of Moral Value," *Ethics*, XLI (1931).
"The Meaning of the Notation of Mathematics and Logic," *Monist*, XLI (1931).
"Kant's Theory of Aesthetics," *Philosophical Review*, XL (1931).
"Esthetics and Epistemology," *Journal of Philosophy*, XXXVI (1939).
"Modern Logic and the Task of the Natural Sciences," *Sigma Xi Quarterly*, XXVIII (1940).
"A Precise Meaning for Objective and Subjective in Value Theory," *Journal of Philosophy*, XXXVII (1940).

"The Use and Abuse of Words," *Journal of Philosophy,* XXXIX (1942).
"Scientific Method and Knowledge," *Philosophy of Science,* X (1943).
"Moral Individualism and Political Value," *Essays on the Theory of Value and Valuation* (Southwestern Philosophical Society, 1945).
"Metaphysics as Hypothesis," *Journal of Philosophy,* XLIV (1947).
"Ethics as Hypothesis," *Philosophical Review,* LVI (1947).
"Reply to Mr. Berenda's Comments on 'Metaphysics as Hypothesis'," *Journal of Philosophy,* XLV (1948).
"Theoretic Knowledge and Hypothesis," *Psychological Review,* LVII (1950).
"Questions of Fact or Questions of Interpretation," *Journal of Philosophy,* XLVIII (1951).
"The Factor of Economic Status in Professional Recruitment," *AAUP Bulletin,* XXXVII (1951).
"A Criticism of the Marxian Interpretation of History," *Tulane Studies in Philosophy,* I (1952).
"What Will Happen in 1965?" *College Public Relations Quarterly,* III (1952).
"Father Parmenides; or Further Concerning Negative Facts," *Journal of Philosophy,* L (1953).
"An Epistemological Analysis of Induction," *Tulane Studies in Philosophy,* II (1953).
"The Hypothetical Nature of Historical Knowledge," *Journal of Philosophy,* LI (1954).
"A Semantic Criticism of Reichenbach's Ethics," *Ethics,* LXIV (1954).
"The Rigidity of Kant's Categories," *Tulane Studies in Philosophy,* III (1954).
"Royce as Logician," *Tulane Studies in Philosophy,* IV (1955).
"Knowledge and Truth in Historical Inquiry," *Tulane Studies in Philosophy,* V (1956).
"Is a Bad Law a Real Law?," *Tulane Law Review,* XXXI (1957).
"What Are So-called Moral Judgments About?," *Tulane Studies in Philosophy,* VI (1957).
"Note on 'Implication' and 'The Assertion Sign' in Whitehead

and Russell's *Principia Mathematica*," *Mind*, LXVII (1958).
"Philosophy and the Categories of Experience," *Tulane Studies in Philosophy*, VII (1958).
"Bergson's Two Ways of Knowing," *Tulane Studies in Philosophy*, VIII (1959).
"Causal Efficiacy and Continuity in Whitehead's Philosophy," *Tulane Studies in Philosophy*, X (1961).
"Mead's Doctrine of the Past," *Tulane Studies in Philosophy*, XII (1963).
"Perception and Epistemology," *Tulane Studies in Philosophy*, XIII (1964).
"Suggestions Toward a Contemporary Epistemology," a series of three articles in the *Southern Journal of Philosophy*, II & III (1964–65).
i. "The Context".
ii. "Fact and Knowledge".
iii. "Theory and Knowledge".
"Are Zeno's Paradoxes Based on a Mistake?", *Mind*, LXXIV (1965).
"A Fitting Theory of Truth," *Tulane Studies in Philosophy*, XIV (1965).
"On the Use and Interpretation of Logical Symbols," *Tulane Studies in Philosophy*, XVI (1967).
"Conceptual Models in Knowledge," *Tulane Studies in Philosophy*, XVII (1968).
Reprint of article "Theoretic Knowledge and Hypothesis" (see above, 1950) in anthology *Educational Research*, edited by Gephart and Ingle (Columbus, Ohio, 1969).
"Two Views of the Nature of Knowledge," *Tulane Studies in Philosophy*, XVIII (1969).
"The Use of Conceptual Models in Understanding," *Southwestern Journal of Philosophy*, I (1970).
"Action, Perception, and Art," *Tulane Studies in Philosophy*, XIX (1970).
"Action, Perception, and Art," (a shorter version of the same paper as above), *Southwestern Journal of Philosophy*, I (1970).

Accepted but not yet published

"Zeno Cannot Be Caught On His Own Racetrack," *Mind*.
Percepts, Concepts, and Theoretic Knowledge, Memphis State University Press.